John Coleman Adams

Nature Studies in Berkshire

John Coleman Adams

Nature Studies in Berkshire

ISBN/EAN: 9783337026288

Printed in Europe, USA, Canada, Australia, Japan

Cover: Foto ©berggeist007 / pixelio.de

More available books at **www.hansebooks.com**

Nature Studies in Berkshire

BY

JOHN COLEMAN ADAMS

Photogravure Edition, with 16 illustrations in photogravure . $4.50

Popular Edition, illustrated . $

"The spirit of the region is very happily caught by the author, who is fond of outdoors, and a sympathetic chronicler of the events of field and woodland. . . . The pictures in the book are very fine indeed. . . . The style of the narrative is clear and unaffected, and the book is one that will not easily be relinquished when once taken in hand. The book is attractive and sumptuous, a credit to the printer's art."—*Chicago Evening Post.*

G. P. PUTNAM'S SONS

NEW YORK AND LONDON

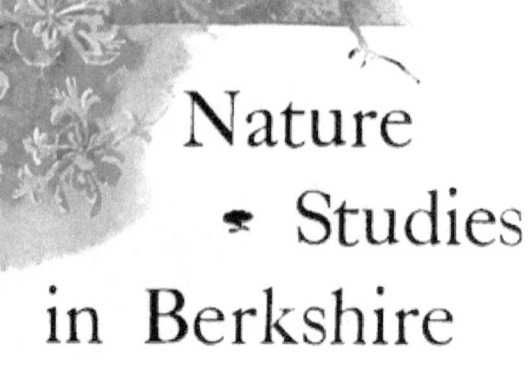

Nature Studies in Berkshire

By
John Coleman Adams

WITH ILLUSTRATIONS FROM ORIGINAL
PHOTOGRAPHS BY

Arthur Scott

POPULAR EDITION

NEW YORK AND LONDON
G. P. PUTNAM'S SONS
The Knickerbocker Press
1901

COPYRIGHT, 1899
BY
JOHN COLEMAN ADAMS
Entered at Stationers' Hall, London
Set up and electrotyped May, 1899.   Reprinted February, 1901.

To My Children

CONTENTS.

CHAPTER.		PAGE.
I.—Our Berkshire		3
II.—Under the Maples		13
III.—A Berkshire Flood		21
IV.—The Dome of the Taconics		31
V.—A Whisper from the Pines		43
VI.—A Hill Pasture		51
VII.—The Circumvention of Greylock		59
VIII.—Berkshire Glimpses		69
IX.—A May-day on Monument		77
X.—Among the Clouds		87
XI.—The Social Flowers		95
XII.—The Berkshire River		105
XIII.—The Epic of the Cornfield		115
XIV.—The Seamy Side of Summer		127
XV.—Fruitful Trees		135
XVI.—The Wings of the Wind		147
XVII.—At the Sign of the Beautiful Star		159
XVIII.—By the Blithe Brook		169
XIX.—The Great Cloud Drive		177
XX.—The Ferns of the Wood		185
XXI.—Living with a Lake		197
XXII.—The Fruitage of Beauty		209
XXIII.—A Quest for Winter		217

ILLUSTRATIONS.

	Page
The Red Bridge, Sheffield . . Frontispiece	
From the Egremont Road.	
Prospect Lake, Egremont	22
From the Hillsdale Road.	
The Dome of the Taconics, Sheffield . . .	28
From the Goodale Quarry.	
Old Pine Tree, Great Barrington . . .	50
Looking toward New Marlboro.	
Mount Greylock, Adams	60
From the "Back Road," Adams.	
Jug End, Egremont	72
Looking toward Norfolk Hills.	
Monument Mountain, Stockbridge . . .	82
Looking across Flooded Fields, Great Barrington.	
Clouds on The Dome, Sheffield	90
Looking south from Egremont.	
A Berkshire Thicket	96
At the Natural Bridge, North Adams.	

Illustrations

	Page
The Harvest-Time, Egremont	116
Looking toward The Dome.	
The Edge of the Storm, Egremont	148
Looking across Fields toward Hillsdale.	
Lenox Road and Stockbridge Bowl, Stockbridge	164
Looking toward Lenox Mountain.	
The Great Cloud Drive, Egremont	178
Across Hills toward Mount Washington.	
The Haunt of the Ferns, Great Barrington	192
In Woods near Hubbard's Brook.	
Lake Pontoosuc, Pittsfield	198
Mount Greylock Range in the distance.	
Williamstown Hills, Williamstown	210
Looking toward North Adams.	

1. OUR BERKSHIRE.

Where rippling fields of wheat and rye
Along the level uplands lie,
And in the valley's cup is born
'Mid rustling green the tasselled corn;
Where ample meadows downward go
To meet the laughing brook below.
. . . . . .
Where stand apart the whispering woods,
Where noises are and solitudes,
Where birds repeat their careless lay
Thro' all the livelong summer day.

Elaine Goodale.

OUR BERKSHIRE.

THE best account of Berkshire must be a record of impressions. No catalogue of places, no word-pictures of its scenery can reproduce the charm its lovers feel. But a recital of what one has enjoyed among these hills and under these skies may recall to others like sensations and delights. This is the only way in which the charm of locality can be passed along. Original enthusiasms make strong second-hand impressions. There is no trouble about the effect of Berkshire at first-hand. It always masters the lover of nature, and makes him an admirer forever, as Scotland does, and Switzerland, and Italy.

To know Berkshire is to love it. To love it is to feel a sort of proprietorship in it, a pride in its glories, a joy in its beauties, such as owners have in their estates, and patriots in their native land. He who was born here, clings to the soil if he stays, or reverts to it if he moves from it, with a New England steadfastness, as intense and deep as a moral principle. He who visits Berkshire is almost certain to visit again and yet once more. He would fain revel in the old delight of air and scene and influence. He believes he has not exhausted the possible experiences to be found in this spot. And so the charm grows, and the sense of belonging to the soil, and the belief that there is nowhere the like of this blend of tonic air and restful scenery, of wild nature and cultivated land, of hill-country and broad plains.

Probably every lover of Berkshire has his own views as to the finest approach to the enchanted country ; and there are many things to be said for every one of them. If one comes, as most outsiders do, from the South, he will have the keenest delight in the winding path which the railway follows along the sinuous Housatonic, and the foot-hills of the Taconic range in north-western Connecticut.

From New Milford to Canaan the route is through a country which only lacks the larger dimensions to be a replica of the Blue Ridge and West Virginia. The railway is a sort of pocket edition of the Baltimore and Ohio. There are the same swift rushes along river banks ; the same wild plunges down grade, and laboured climbing up again; the same sharp curves around rocky corners, where the steep slopes come close to the rapid current ; the same forest-clad ridges ; the same dark-green ravines along whose lowest levels the stream slides downward toward the sea. One sees, too, among his fellow-passengers the same drawn lips, the same ashy pallor, that betray the "train-sickness" which is but little less than seasickness.

And when at last the valley broadens above Falls

Village, and across the plains of Canaan, to the westward, the blue shoulders of the Dome and Race Mountain swell against the sky, the traveller feels that he has been ushered fittingly into this fair hill-country.

Yet from the westward the approach is no less fascinating, perhaps is more impressive. There is the long range of the Taconics, stretching from Williamstown to Boston Corners, a continuous rampart, a natural boundary line between the states whose borders meet along its ridges. Coming toward it from the Hudson, one sees its surfaces, unbroken at first, and reared like a solid barrier, by-and-by divide into separate hills, drop away into ravines, open a way here and there for a winding road, through passes and valleys and notches. Then to climb the long steep slopes, and twist around the mountain-sides, and from the broad ridges look down into the valleys and across to the innumerable hills of Berkshire, is to be filled with all the joy of a sudden and ecstatic vision.

There is a peculiar romance about the Williamstown portal to the county, for that was once the gateway by which the old Indian trail, entering from the westward, and trodden by many a war-party and many a squad of dusky hunters, led down to the Connecticut and Housatonic valleys. One can hardly commend the eastern gateways to the enchanted country. For though they are rugged and wild and picturesque they deny to the wayfarer that definite sense of entrance which so much adds to the enjoyment of coming to a famous region.

Perhaps if suddenness and surprise were what are sought, the sensation might be depended on if one entered by way of Hoosac Tunnel. But there are drawbacks to the experience of being hurled at grand and imposing scenery out of a hole.

There is a special attraction, in coming to Berkshire from the south, in the keen perception of the changing air, the quickening of the pulse under the gentle tonic of the climate. There are few finer experiences to be crowded into a few brief hours, than the change in midsummer from a New York morning to a Berkshire noon. One leaves the stifling city, steaming in a dog-day fog, sultry, humid, lifeless, reeking with acrid odours, heavy with smoke, and gas, and dust. He swelters and suffocates in the great iron half-barrel of the Grand Central Station. He gasps away his few remaining breaths in the tight-shut cars rattling through the dark tunnel.

Just before the traveller is dead he reaches open air and daylight in Harlem, and sniffs the salty breath of the Sound and the sedges along the shore. Then comes the northward turn at South Norwalk; and still the thick air is laden with vapours and the languid lungs cannot get oxygen enough to feed the fires of life. But now at least there is no more gas to drug the ozone; and by the time Bethel is passed the odours of the field replace the clogging fumes of brewery and mill; and the other side of Danbury the steam dies out and leaves an atmosphere which but for the cindery breath of the engine would be ninety-five per cent. pure. But the real change, the awakening to the consciousness of a new vitality in every draught, the sense of the gentle tonic of the Berkshire Hills, only begins when Brookfield Junction is passed, and New Milford, and down from the slopes of the Litchfield Hills, and out of the valleys of Cornwall and Canaan, come the sweet and quickening airs which put life where languor was, and stir a quicker pulse in the weariest heart.

It is hard to do justice to this climatic charm, without seeming to drop into unqualified panegyric; in which attitude one is immediately suspected of special pleading and remunerated enthusiasm. But the truth must be told about the air of Berkshire, even at the risk of seeming partisanship and an excess of adjectives. It comes near to being ideal. Yet it is more than likely that the stranger will begin by thinking it not so very different from any climate he has been used to. It is tonic. But the stimulation it imparts is not like that which comes from the wine when it is red; it is rather like the draught at the wayside spring to the hot and thirsty wayfarer. It is not aggressive; yet it accomplishes things. Tired people do not realise that it is affecting them until, after a fortnight of it, they begin to wake up in the morning without any tired feeling.

There are "bracing" airs which are almost fierce in their attack upon an enfeebled or tired system,

They come at one like a man with a club. As one breathes them he is reminded of an old toper's dram which "bites all the way down." But the Berkshire climate does nothing like this. Its touch is gentle, its manner mild. It does good in a way one would almost call insidious. It banishes languor without creating an unwholesome fever for work; and it makes rest refreshing, without adding to the need of it by its own lifelessness. In a word, it is very likely to be just what everybody thought it was not, yet is almost certain to be what the majority of people are glad to have it.

The things here set down are true of Berkshire at any and at all seasons. Its attractiveness is not a remittent trait. Its charm begins in the spring and lasts to the winter's end. For those who love a real winter, with crisp, frosty airs, roaring winds, snow-falls and icy ponds, the chime of the sleigh-bells and the musical crunch of footsteps on the trodden snow, Berkshire will always be fascinating. And when the marsh-marigold in the meadow and the arbutus in the wood blazon the advent of spring, Berkshire but turns a new page in the volume of its delights.

The summers of Berkshire have become the annual necessity of thousands who out of baking cities and their brick and brown-stone caves swarm over its green fields, and sit under its elms and maples, thankful for its fair vistas, its thrifty fields, its noble hills and mountains. And when autumn sets the foliage ablaze, and the hillsides glow and flicker and flame through the haze of October, the happy souls who are permitted to linger or live where they can look upon it all, are filled with fresh wonder that one small county can so vary and renew its charm and be always in its best season.

Our Berkshire has another peculiarity all its own. It has a triune existence. It is three counties in one; and he who knows the one county may have little or no knowledge of the other two. There is the commercial Berkshire, the Berkshire of the farmer and the manufacturer and the merchant and, perhaps one ought to add, the hotel-keeper. And this is a county important and impressive. These smooth meadows, these uplands and vales, are not mere lawns, kept trim and neat for the benefit of visitors and idlers by the way. They bear real harvests. They yield a revenue to the purses of their owners. These busy brooks are broken to harness and forced to furnish power and privileges to clanging mills. There is an industrial Berkshire built on the ruins of two or three earlier periods, bearing lively witness to the thrift, the energy, the courage of the business men of this county. And he who comes to see this Berkshire may know nothing at all of the other two.

For the social Berkshire which has become so famous in late years is a world by itself. It migrates from the cities in the spring, and it flies southward when the leaves fall and the frosts bite, just as the birds do in the forests round about Lenox and Stockbridge and Pittsfield. It recks little of the prosaic life which thrives the year round in the county. But it has a life of its own, brilliant, charming, intoxicating, a round of interchanges of courtesy, of fêtes and festivities, of golfing and coaching and dining.

There is another Berkshire still, the background of these other two, on which they flit and flutter in their changes and chances, as clouds come and go against the eternal firmament. It is the Berkshire which nature has fashioned, enthroned on the hills, ensconced in the valleys, singing in the brooks, waving hands of welcome and farewell from the branching trees, weaving rich patterns on the turf with wild flower, and grass, and vine, and shrub. That, after all, is the real Berkshire, the one they love the most who love Berkshire for its own sake and for nature's sake. It is " our " Berkshire, whosoever we may be that love the open sky, and the forest glades, and the hilltops and the brooks. It is such a region as Stevenson celebrates in those thrilling lines,—

"It's ill to loose the bonds that God ordained to bind.
Still will we be the children of the heather and the wind;
Far away from home, O it's still for you and me
That the broom is blooming bonnie in the north countree."

2. UNDER THE MAPLES.

Nor unremembered here the garish stage
   Nor the wild city's uproar, nor the race
For gain and power in which we all engage;
   But here remembered dimly in a dream,
As something fretful that has ceased to fret.
WILLIAM WINTER.

UNDER THE MAPLES.

AT last the dream of many weeks is realised. The hot and steaming city is leagues away. Its rattle and its racket are become a mere memory of pain. The eye no longer wearies of the endless procession of men and women who defile from morning till night over the pavements. The wandering glance no longer rests on tin-roofs baking in the uncompromising glare of the sun, or roves down into backyards which seem like so many pens for catching the heat. All that is vanished; and instead of it, a scene meets the eye in which one loses sense and thought in a sweet oblivion of content. To be sure it is hot; because this is July and heat is to be expected. The air quivers and throbs over the rye-field. The far hills retreat still farther behind a blue haze. Overhead the clear azure of the sky deepens to an intenser shade as one sees it against the bright green of the maples. The maples themselves arch the table which serves for a summer writing-desk with a thick shade, and the fresh breeze which cools the spaces under the branches voices itself in a cheery song among the leaves. Under the maples here in Berkshire is an incomparable vantage-ground from which to behold the glories of midsummer as they pass by.

For consider the maples themselves. It is a most distinguished group which surrounds our home. The house itself stands full and clear in the sunshine, accessible to light and air on every side. But within a rod of its walls, like a great green colonnade about this domestic basilica, these thrifty maples lift their engirdling foliage. On two sides a double row of them forms a cloister around the little lawn, under whose leafy roof one may walk at noonday with scarcely a patch of sunlight falling upon him. These two trees at the corner of the colonnade are conspicuous in this regard. They have grown like twin brothers. Their inner boughs are linked and twined together. They are healthy and affluent and strong. As one lies on the turf beneath them and tries to find the sky through their branches, he sees layer after layer of spreading leaves, like an ample thatch, a protection alike from sun and from rain. All day one may sit under this roof of green, and be perfectly secure from any prying ray of sunshine; and when the rain falls it is sometimes hours before the drip from leaf to leaf reaches the turf and moistens the grass about these sturdy trunks. But under their broad branches the breeze moves, unstayed and free, tempering the fierce heat of July noons, and bringing renewal of life with the midsummer twilight.

It moves through the foliage, too, with a voice all its own. It may not be that every tree speaks with an individual tone when the wind breathes through it. But there is a difference in the voices of the trees which even the least experienced dweller beneath them must observe. If the pines utter a deep contralto note, full of pathos and suggestions of the undying solemnities of the world, surely the voice of the maples is the stirring tenor, breathing the lively song of action, the chant of good cheer, and the prophecy of weal. The note of the pines is the murmur of the sea repeating itself in the depths of the forest. The note of the maple comes nearer to the blending sounds of a great city, where human life surges and breaks upon the pavements. It seems as if the maples had imbibed something of the life and spirit of that race by whose homes they have grown lo, these many ages, and as if they wafted back to the heart of the listener who stands beneath them the chorused voices of his own deep thoughts, his strong impulses, his vigorous ambitions. To be sure the maple is still associated with the forest, and so far has a suggestion of wild nature. But seen near the homes of man, sheltering his roof and shading his dooryard, it sheds its wildness as the house-dog sheds his wolfish traits.

Perhaps it may seem as if with such an impression of the maples one might not find them restful trees, nor draw from them the sweet anodyne of oblivion to the year's busy days and works. But the suggestions of human life which rustle in these leaf-voices, the echo of one's activities of brain and heart which whisper or roar from these shrill chords, are all on its poetic side; and it is good to have one's life sung over again in any music of nature or of art. The poet who unbraids the strand of romance from the dullest life, and weaves it into his song, soothes as he sings the heart of him who bows under the burden which the poem celebrates. A man's home never suggests his daily drudgeries when it looks back at him from the canvas of the artist. These maple-trees give back such a dreamy, idealised echo of humanity and its stir and bustle and business, that one takes it as he does the song which repeats his sorrow and the picture which shows his home. One lies beneath them and looks out upon the fields, as he would lie upon the safe cliff and behold the sea break impotently at his feet. They suggest the world, but they breathe forgetfulness and comfort, too. They are friendly in their hints, and only stir the happiest memories.

But the maples are not always moved to speech. There are hours in the morning or late in the hot afternoons when their leaves are as motionless as the everlasting hills. There are moments when leaves, and twigs, and the smaller boughs rise and fall with a movement as faint as the last breath of old age. These are the times when they lull to dreamless sleep, or to the dreams which do not wait on sleep. Then the drone of the locust comes up from the field; the melodies of the small birds beguile the ear; and one almost fancies he can hear the corn growing and the sap flowing in the maple-tree's veins.

Happy those mortals who can build their homes, and do their work, and think their thoughts, and bear the great experiences of life under such shades as these, which bring something of the voices and the spirit of the forest shades, down to the roadside and close to the cottage threshold! Thrice blessed the man who beneath their gracious arms finds rest, healing, and new courage after the battle of a year's toil.

3. A BERKSHIRE FLOOD.

This feast-day of the sun, his altar there
In the broad west has blazed for vesper-song.
                              DANTE G. ROSSETTI.

A BERKSHIRE FLOOD.

THE sun had been shining in a clear sky for three days, and now, as the meridian of the third day was passed, I could no longer resist a conviction which had haunted and made me restless many times, that yonder, in the valley behind the "Jug End" range, there were lights and shades awaiting the eye which one could ill afford to lose. So in entire disregard of the intimations that it was too hot for walking purposes, I climbed the rail-fence behind the barn and started over Pasture Hill.

There is something about such an August afternoon as this was, which makes one understand and respect the motive and the method of the impressionists in painting. It was one of those days when the one overpowering, dominant, and irresistible thing in the landscape is light, the palpitating, luminous atmosphere, crowding in before everything of substance and of form, itself almost becoming substantial and taking form. The world is drenched with sunshine. The air is saturated with this downpour of sunbeams. Just as the raindrops glance from the rocks and ledges, and gather in trickling streams, and collect in the deeps of hollow pools, so the light seems to slide off from every surface and substance in the landscape, and stream back into space, to ripple and flow, and stand, deep and liquid as that other element, in the deep places of the atmosphere. One realises now the full meaning of the phrase which speaks of a "flood" of light. This is a veritable luminous deluge. Given forty days of such a downpour, even with the intermitting nights to break the effect, and the world would sink beneath these waves of sunshine, as in a great deep.

That was the first feeling which was floated back from the scene which opened from the top of little Pasture Hill. Every distant hill and mountain was swimming in light. Every near field and wood was afloat in the same eddying stream. Not since the "Fiat lux" was pronounced has the earth weltered in a more unfathomed sea of sunshine. One could forgive the impressionists many a vagary, many an extravagance of purple and violet and lavender, many a suppression of other truths about nature in her outdoor moods, for the sake of their fidelity and truth in emphasising the aggressiveness of light and the force with which it sometimes thrusts itself upon the sense as a magnificent fact of the landscape, worthy of a thought in and of itself.

But one may not pause on the hilltop. The valley tempts the vagrant feet on over the two fences to the fair slope of a cleared hill beyond. It is a favourite stroll and never so enticing as when this August sunlight glorifies and transfigures it. The

Prospect Lake, Egremont.
(From the Hillsdale Road.)
" *The world is drenched with sunshine.* "

path led around a wooded swamp, within whose shadows, as the afternoon waned, a multitude of bird voices would blend in a vesper carol. But now only one harsh cry was to be heard. I am a stranger in bird-land, but from the strident quality of that call I imagined it to be a bird of prey, a hawk perchance, or some other predatory creature, and wondered what social or domestic tragedy it betokened. But aside from this discordant note, all was serene and restful, in sound as well as in sight.

On the slopes of the upland my way led among the slender spikes of the blue vervain and the bushes which gleamed with the golden blossoms of the shrubby cinquefoil. These neighbours of the field have a curious habit which reminds one of the practices of poor Mr. Wilfer who never was able to afford enough money at any one time for a new suit of clothes, but must wear his hat quite shabby before he could get a new coat and whose coat was threadbare before he could command new shoes. The shrubby cinquefoil and the vervain never seem able to muster sufficient resources to blossom all at once. They do their flowering a little at a time, and so contrive always to look shabby even while they are doing their best to make a good appearance. One can hardly imagine the splendour of purple and gold which would blaze in this hillside field, if these neighbouring wild-flowers should become suddenly rich, and bloom all over at once, like the golden-rod or the laurel.

Crossing the ridge and the valley on its western side, and extricating myself from the threatening complications of a barbed-wire fence, I found myself at the entrance of what was once a town road, now disused on account of a better one which had been made about half a mile to the north. There is always something a little pathetic about an abandoned road, growing up with weeds and vines and grasses, a memory of usefulness past, a picturesque ruin for the present. But there is really no need of wasting sentiment on it. The abandoned road generally means that a better one has taken its place. The wild things are growing in the old road because travel is smoother, easier, and safer in the new. The abandoned farms and homesteads, too, in many New England towns, most frequently tell a tale not of ruin to the household, but of a larger prosperity or a better chance for it which has taken them to fairer fields. I was willing to accept the old road as a decorative effect in the landscape, and let it pass for that. It was a good short cut to the glen I was seeking, and it afforded a most entrancing view of a wooded range of mountains just catching the shadows as the sun slanted across the topmost ridges.

This point of view was another witness in favour of the impressionists. The scene it commanded bore ample testimony in favour of their favourite purples and blues. This early twilight on the shady side of the hills is a rapture to the lover of colour. If nature could always have the "blues" like this nobody would complain. One learns to love it and to crave it, and to take it as an intoxicant to the eyes. To-day it was one shifting mass of hues and tints, always with blue as the key-note, but running over the whole gamut of shades. Nevertheless, the whole of this landscape could never have been translated to canvas by means of indigo and its kindred colours. For glorious greens were here, no longer, indeed, the fresh emeralds of spring or early summer, but mellowed into browner and yellower shades, the riper draperies of autumnal days. Here, too, were the pervasive yellows which flow with the sunshine and ebb back from leaf of tree and grass of field, from wild-flowers by the wayside and ripening grain upon a score of hills. I do not know who is responsible for the assertion I have heard recently that nature is sparing of yellows, but it is certainly a grave misrepresentation ; and it would be by no means strange if a whole school of artists should presently arise to tell us that there is yellow in nature, and do it with such force and pertinacity as to make us believe that Mother Earth is suffering from the jaundice.

But now I had reached the mouth of the deep glen whither I had bent my steps, just in time for the finest hour of the day. The sinking sun was lost already to the eastern slopes, and their deep chestnut forests were fast gathering the deeper shadows. Eastward the beetling rocks and splintery woods still held the sunshine. Three miles away, at the head of the lovely vale, the graceful Dome of the Taconics swelled into the full light, against a clear blue firmament. The farm-lands, cleared well up the mountain-sides, looked rich and thrifty. Over the whole scene brooded an air of serenity and quiet such as Millet has put into the famous "Angelus." The light which fell down from the hills and filled the deep glen was that tender, pensive glow in which is blended the spirit of two afternoons,—the afternoon of the day and of the summer.

But the fulness of the flood was past. There was a subsidence of the deep light in the sky above and a fading of the glow upon the earth beneath. Slowly the torrents of sunbeams, which all day had poured from the golden urn, the sun's exhaustless reservoir, began to slacken in their flow. Then, as always with a great deluge when the rains have ceased and the feeding streams, there began the slow ebb of the mighty tides. Down the slopes, yard by yard and rod by rod, out of the open glen, over the wide rolling meadows, the great August freshet subsides. The sunshine rolls backward in great surges and little waves toward the west. The outlines of the mountains and the forests and the rocks grow hard and dark, like land newly emerging from the deep. The glory fades from the landscape. Up from deeps of the glen come the lowing of kine, the shout of the herdsmen, as they follow the retreating tides and return to their homes, as men come back to houses whence the rising waters drove them out. From the forest-shadows and from the boughs of orchard and roadside trees comes an outburst of song by a great multitude of birds, such as might have swelled about the ark when its windows were opened and the songsters released from their long imprisonment. The cheerful robin warbles his content, the song-sparrow adds his happy measure, and at intervals there comes from afar the solemn voice of the thrush, like the intoning of an evening prayer. But soon a hush falls upon forest and field.

> "Now fades the glimmering landscape on the sight
> And all the air a solemn stillness holds."

The Berkshire flood is over and its tides go down with the sun.

The Dome of the Taconics, Sheffield.
(From the Goodale Quarry.)
"*A mountain of no mean proportions, walling up the western view.*"

4. THE DOME OF THE TACONICS.

Pillar which God aloft had set,
So that men might it not forget ;
It should be their life's ornament
And mix itself with each event ;
Gauge and calendar and dial,
Weather-glass and chemic phial,
Garden of berries, perch of birds,
Pasture of pool-haunting herds,
Graced by each change of sun untold,
Earth-baking heat, stone-cleaving cold.

R. W. EMERSON.

THE DOME OF THE TACONICS.

THE traveller by the Housatonic Railway, northward bound from New York, becomes aware, soon after leaving Ashley Falls, of a mountain of no mean proportions, walling up the western view. It is a mountain whose smooth outlines and gracious curves somewhat disguise its real proportions. At first sight the supercilious tourist from more ambitious altitudes might easily pass by its beauties and count it a small affair; though his map will tell him that it is a bit higher than the Mount Desert hills, and the peer of many of the lesser peaks in the White Mountain ranges.

But it is never safe to despise a mountain because it does not stun you with its proportions. The man who respects a high peak solely because of the challenge it makes to his eye or to his muscles, may be an athlete or a sensation-seeker; he is not a true lover of the high places of earth. Altitude is only an incident of a mountain. It is not essential either to its beauty or to its impressiveness. I can easily imagine that a sensitive and self-respecting peak might be as unwilling to be judged by its height alone as a man would be. Only the grandeurs and sublimities of the earth are emphasised by enormous heights. Its beauties lie closest to the less ambitious hills.

So, as the eye becomes more familiar with this peak which stands guard over Sheffield and Egremont, one learns to hold it in greater respect and affection. Seen from the South Egremont and Great Barrington side, whence it seems to rise abruptly from the valley, and make first a quick and then a more gradual approach to the clouds, it justifies the preference of the older people of the region, who resented the attempt to christen it "Mt. Everett," and clung to the old name, "The Dome." A dome it is in outline and in mass, rounded, soaring, crowning its lower spurs and ridges, as the grand curves of St. Peter's top its nave and transepts. It rises twenty-six hundred feet above sea-level, and nineteen hundred above the valley. Its sides are clad in a growth of maples, chestnuts, and birches, as far as the upper ledges where the scrub-oaks and pines compete with the blueberry bushes in the struggle for existence. Here and there a few pines and hemlocks remain to tell the tale of a glory which has been shorn away from these slopes; and in the very heart of the mountainside, facing the sunrise, a deep and precipitous glen, the channel of a fitful stream, is dark with evergreen foliage.

Of course the summit of this mountain early became a place coveted for the soles of my feet. There is but little pedestrian ambition among the Berkshire

The Dome of the Taconics.

summer population, especially if the high hills are the objective points. The usual method of getting up the Dome from South Egremont, is to ride some ten miles into the town of Mount Washington, to a point within less than twenty minutes from the summit; and so all the delights of the ascent are diluted into a kind of waggon picnic. Nobody seemed to know of any good path from the South Egremont or eastern side; and it took a little faith and common sense to decide that the climb could be made from this side under much better conditions than by the roundabout and lazy route most in vogue.

Early in the season a little party of us undertook the ascent, found an easy and romantic path, and enjoyed a glorious view. Being sure of the pleasures of this trip, I determined to summon my comrade in spring outings, himself a tall dominie, and introduce him to the beauties of his own contiguous territory, the fair valleys and hills of Columbia County, New York. He was entirely willing to take a small holiday and answered the call with alacrity, bringing fair weather and a northerly breeze along with him. Monday morning brought him to us apparently as fresh and bright as if it were not the day after Sunday, and by two o'clock we were bowling along in our farmer's "two-seater"—an instructor in New York University, a freshman in the same college, the other dominie, and myself. We survived the suffocating dust which rose from the parched roads in the four-mile drive to the Knickerbocker farm, and at two twenty-five slung our lunch-boxes and started up the path for the summit. Like most well-regulated mountain trails this one began in a wood-road, old and grass-grown and mossy, but easy of grade and for the most part clear to the eye.

The first half-hour was spent in doubling backward and forward, beating up-hill as it were, against a fairly heavy grade. The way was lined with hardwood bushes, with ferns and mosses most cool and refreshing to sight and to smell. Here and there a group of wild sunflowers lent their bright yellow hues to the scene, and once we stumbled upon some foxglove. The mountain vegetation was showing the effects of the drought which had been searing the fields in the valleys, and had made great strides toward autumnal hues in the three weeks since we were over the path before. A little spring beside which we had eaten our supper on that previous ascent was now but the shadow of a refreshing name, a mere dry and empty earthen bowl. We were disappointed in our expectations of a cool draught at this wayside fountain, but remembering another, only a mile or so farther up, we pushed on in hope.

Ten minutes more along a pathway through fresh young birches and maples, brought us to the turning in the path where before we had made the mistake of keeping straight on which cost us an hour and a half of needless tramping. This time we swung to the left, by the "blazes" we ourselves had made, and came suddenly into the finest stretch of the whole route, a path lined for nearly a mile with the dark, glossy leaves of the mountain laurel; a path which was glorious for refreshing, deep, restful green in this hot August day; a path which in June, with its blossoms rivalling the pink and white of the apple-trees in the plains below, must be more magnificent than the costliest gardens of royalty. Nor are August and June the only months in which this path would woo the lover of the beautiful; for all along its sides, peeping out from beneath the underbrush and the creeping vines, were the unmistakable leaves and hairy stems of the trailing arbutus, where next May the tiny pink cups of its blossoms will be uplifted, full to the brim of the choicest fragrance of the spring.

Soon after the laurels are passed, the path crosses the bed of a stream, and turns a sharp corner to the right, close by an old corduroy bridge. A few rods to the east of it is a spring whose waters we had found as abundant as they were sweet and cool, but which now was dried up to about a scanty pint, lingering in the last hollows of its rocks. But there was enough to slake our moderate thirst and leave sufficient for our return. Then for a half-mile we hastened southwest through an almost level path which winds about the base of the mountain's last great dome, before we bent sharply to the south-east and scrambled up the narrow footpath toward the open ledges. The sun came slanting through the saplings which crowd close to the path, slender shafts of "splintered sunlight" which fall most temperately into these cool shades, through which we clambered at a round pace, which soon brought us to the pole that marks the highest point upon the Dome.

It was a noble prospect which opened all around us. From this extreme south-western corner of Massachusetts the hills and vales of four States stretch away to distant horizons. Northward, beyond the saddle-back of Greylock, lay the lower hills of Vermont. Eastward were the hill-towns of Massachusetts, Berkshire and Hampden farms and villages. To the south the Taconic range strayed off into Connecticut and lost itself, and westward the valley of the Hudson swept broad and green up to the base of the Catskills. It was a scene to charm and rest the eye and mind. It was an outlook upon nature not in her moods of wildness or of solitude, but as she comes from the hand of man, trimmed of her asperities, combed and brushed with axe and plough, with a gracious air of cultivation and of refinement.

Everywhere about the mountain's base, far as the eye could reach, were the splendid farms of Berkshire and of Columbia. On the east the valley of the Housatonic; westward Green River and its smooth meadows and low hills. It was a surprise and a pleasure to note how much forest is still left on these mountains and hills, nor can one fail to connect the verdure of these countless acres of farming land with the other acres of wooded heights. The

The Dome of the Taconics.

Catskills were immersed in a blue haze whose tremulous lights hid all but the faintest outlines of those romantic peaks ; while faint and far, a mere hint of a range in the distance, we could make out the Shawangunk Mountains, flanked on the east with the Highlands of the Hudson.

The impression left upon my mind after many trips to this delightful summit is a strong feeling of its resemblance to the Lake Country in England. One has the same sense of wildness which has been treated and reduced by man's care. There is the same mingling of mountainside and farm-meadow. If the American scene is lacking in the lakes, the water glimpses which make the English one so fair, it is superior in the splendid sweep of the landscape lines, and in the ample forests on these great hill ranges. Bryant puts the double charm of these Berkshire views in his lines :

> " Thou shalt gaze at once
> Here on white villages and tilth and herds,
> And swarming roads, and there on solitudes
> That only hear the torrent, and the wind,
> And eagle's shriek."

But the declining sun warned us that we must turn valleyward again ; and with reluctant feet, as every lover of the heights must leave them, we plunged into the gathering shadows of the woods. We made quick time back to the beginning of the laurel path. But when we turned this woodland corner, a new idea possessed us. We knew we were at the head of the ravine which makes up the mountainside, for we were crossing the dry bed of the brook, which could have no other outlet. Why should we not shorten our way and extend our information by a short cut to the base of the mountains through this "Mossy Gill?" The motion was put to vote and the vote was at once put into motion. We threaded our way along the almost dry bed of the stream, down a grade which at first was adapted to a dignified and graceful gait, as we stepped from stone to stone.

But presently scene, surroundings, and footpath all changed. The young maples and birches gave place to tall hemlocks, looking down on the prostrate trunks of their ancient comrades. The bed of the brook took a more decided pitch downward and the little ledges, worn by the waters of many spring freshets and smooth and slippery with moss, began to make drops of four, and six, and eight feet. We left the boulders and crept along the side of the gorge, thrusting our feet deep into the mouldy soil, and between fallen trunks and tree limbs. Soon we came to a ledge some fifteen feet in height, where a smooth log, caught in the clefts of the rocks, tempted the clerical brother to try a sort of primitive toboggan slide. The slide was accomplished, but the damage to his apparel, especially his pantaloons, was irreparable. The rest of the party slid, crawled, and dropped in other ways to the ledges below.

But a harder pinch came a few moments later, when we found ourselves caught between flanking precipices, sheer and perpendicular and dropping forty or fifty feet, and only the narrow ravine of the brook, and its shrubby, shelving ledges, to offer us a way down. How we got out of the pocket I never shall quite realise. But like many another hard thing in life, we got through before we knew it, and landed fully thirty feet below the edge of the fall, in only two "drops," with no bones broken and only one big rent in the company. Then the path grew easier, and we had time to observe more appreciatively the charming scene. It is a marvel that the place is not better known and more frequently visited by the summer populace of Sheffield and Egremont. If such a wild and romantic glen were in the White Mountains or the Catskills, so rugged, so moist and cool, so upholstered with moss and fern, it would be sought from afar, and the tourist would scatter the shells of his hard-boiled-egg lunches on every stump and boulder. But it is just as well that the possibilities of this great ravine in the Dome should not be sought out. Its beauties will remain the choice possession of the few who have seen and loved them.

We sat us down beside the relics of the brook and ate our supper in the forest twilight, to the sweet accompaniment of a tinkling little rill, all that the drought had spared of the sturdy stream. Then, when we had carefully draped and repaired the tall dominie so that he might venture again among his kind, we trudged through the dust and the dusk a tedious five miles to the cottage.

5. A WHISPER FROM THE PINES.

I stood as still as the solemn firs, and hearkened with waiting mind.

Then I heard far away in the topmost boughs the eternal sough of the wind.

Fraser's Magazine.

A WHISPER FROM THE PINES.

THE great storm which swept by us yesterday, lashing the land with its double scourge of wind and rain, has disappeared in the east. The sky is clear of clouds, and instead of the harsh east wind a fresh breeze from the west is drying out the soaked and sodden earth. No spot on all the farm seems so attractive to-day as these pine-woods. They are the peculiar charm and attraction of our neighbourhood. They have become a favourite spot for writing and for study, and many a page of the summer's work will call up, in the review, the sighing of the wind in the tufted boughs and the glint of the sun on the pine-needles.

To-day I have been studying the grove. Somehow the printed page has lost its fascination, and for the hour, my mind will run along the lines of nature's book, this volume known and read of all men who will take the trouble to open their eyes. And after my readings to-day I confess to a stronger feeling than ever before possessed me, of kinship to these trees, and a blood-relationship running back through countless ages. It is amazing how many of our family traits are the common inheritance from a past which is inconceivably remote; so our behaviour to-day cannot fail to impress the close observer of family likenesses, with our resemblance to our cousins, the trees.

Look at these pines, for example, tall, straight fellows, most of them, running up from forty to sixty feet, rough-coated at the base, but with smooth, sleek jackets toward their tops, which well become their slim, aristocratic figures. Erect and graceful in mien, soft-voiced in the gentle winds, well-clad and well-fed in appearance, they are excellent types of a well-descended and well-connected family, proud of their birth, blood, and breeding. And in their isolation here on the little knoll where they live, forty or fifty of them, one is reminded of the exclusiveness of some human societies, whose aristocratic ways are not unlike those of the lordly pines. But it was only to-day that I discovered how much deeper and farther this analogy ran.

Within a rod of me, as I write, is a pathetic bit of tragedy whose meaning has but just penetrated my mind. Right here, in this group of stately forest aristocrats, has grown up a little, rough-barked, scrubby "pitch-pine" tree. How it happens to be here, the sole one of its kind in the grove, is more than I can tell. But it has rooted itself and grown to a height of nearly forty feet. Yet it could not hold its own. It is among competitors which altogether outclass it. These elegant members of the pine-trees' "best society" have simply crowded out their poor relation. They have squeezed in front of him and taken his light and heat. They have out-climbed and over-shadowed him. In whatever ways trees have of accomplishing the socially murderous feat, they have frozen him out.

The pitch-pine is dead. He has succumbed to the chill of good society. He has gratified his ambition, perhaps, to live the life of the "first families," and the result is, that he amounts to nothing now, save perhaps a cord-foot or so of firewood. I have a feeling that he may have found the living a little too high for him. If he had only stuck to the sand barrens he might have been a stout and thrifty tree. But the good living of this richer soil has taken the energy all out of him ; so that his haughty neighbours have had an easy and a short task in crowding him to the wall. How very human these pine-trees are! Or shall we say how very like the trees these mortals be? It is sometimes a little hard to know which way to put the analogy,—whether to give precedence to the man by virtue of his superiority, or to the tree on account of its seniority. In any country but America genealogy has the preference. In almost any court of Europe the pine-tree would go in to dinner before the human.

The whole grove is a witness to this exclusiveness and selfishness of the pines. The little group has but scant underbrush, even where it has been allowed to grow ; and within short range of my eye there are half a dozen other groups of pines whence other trees have been quite excluded. The pines stand by one another. They crowd close together, and make common cause against intruders and strangers to their set. They mass their evergreen foliage in a thatch which casts a perennial shadow on the ground beneath, very discouraging to shrubs and trees which love their share of sunlight. And then they cover the ground with a carpet of rusty, cast-off needles, which seems still more to disconcert the grass and vines and creeping things. Give these pines any sort of a hold and they will maintain it against all comers. The adults of the family have no bowels of mercy.

But just look down the slope yonder at that little copse of young trees under the ledge. Day before yesterday I counted in that little thicket nine different varieties of trees,—oak, pine, birch, sumac, wild cherry, elm, willow, poplar, and ash. They are getting on well enough together. It is a very democratic group. These juvenile trees play very harmoniously in company. The question of precedence, of rank, of rights and preferments, does not seem to have come up as yet. Indeed there is a pair of young poplars, or "popples" as the boys call them, which seem to have quite taken the lead and to be the chiefs in that little group. But these equalities in the democracy of childhood and youth cannot last; they wane with growth and the maturing of character and its assertion of traits which line men—and trees as well—into groups. Happy childhood, in which caste spirit plays so little part. How strange that it should be so much the same among the children of men and the children of the trees.

But I seem to have been writing down the pines, and giving them a bad character. I suspect I have been caught in the breath of the "spirit of the age," and blown by it quite off the course of my real sentiments. It is the defect of the age that we are envious of the strong; and we find fault with them whenever they are even unconsciously the cause of distress or disaster to the weak who get in the way of their strength, and suffer thereby and therefrom. Ought we, after all, to blame our cousins, the pines, because in being true to themselves, and living strictly and sternly by the law of their family life, they make an environment uncomfortable and impossible for the tree which thrives on a different diet? Shall we condemn the man who has attained a dinner of five courses because his diet and his demeanour are distasteful to his fellow-man who dines sumptuously on black bread and bologna? For myself I hasten to repudiate the levelling sentiments which would put me in the attitude of a critic upon the life of the pine-tree. I believe in the pines. I like their family unity. I honour them for their ambition. I do not blame them because they like their own kind best. I admire the way in which they sturdily push for the top. Above all I confess my debt to them for a lesson of deep and vital spiritual import. Whoever sits here in this grove must see how resolutely these pines push out for the sunshine and the light. Here is one just in front of me, on the edge of the grove facing the morning sun. It is fully seventy feet high. For two-thirds of its height it has not a single limb on the inner or shaded side of its trunk. Until it begins to get the light from the west, above the tops of its companions, it stretches its long branches, like so many outreaching, uplifted, and imploring hands, toward the quarter whence the most light comes. Here are other trees, whose first forty feet are marked by dead and dying branches, dropping into decay because they are starving for light. But this tree will not starve. It pushes its long trunk upward. It reaches after the upper light ; and if it can reach the sunbeams at the top it minds nothing about the shadows below. As long as it keeps in upper sunshine, the bare trunk in the sombre light of the grove, the very roots deep in the dark of the earth beneath, feel the thrill and currents born in those life-giving rays. It does not seem as if any enlargement upon that fact could be more forceful than the mere statement thereof. Long, long ago the ancestral pines began to grow as one day the human soul was to find and keep its life. Happy the man who still holds fast to the ancient and unchanging law.

6. A HILL PASTURE.

The sun looks on our cultivated fields and on the prairies and forests without distinction. They all reflect and absorb his rays alike, and the former make but a small part of the glorious picture he beholds in his daily course. In his view the earth is all equally cultivated like a garden.

Henry D. Thoreau.

Old Pine Tree, Great Barrington.

(Looking toward **New** Marlboro.)

"*Down past the old battered pine and by the edge of the swamp.*"

A HILL PASTURE.

THROUGH the barnyard and over the worn grass of Pasture Hill, down past the corner where the lady-fern and dicksonia grow, under the old, battered pine, and around the northwesterly edge of the swamp,—that is the way to the hill pasture. For this route brings you up to the place where the rail fence is easiest to climb and where moth-mulleins stand guard like tall, slim sentries over the gateway to the field.

As for times and seasons, it is best to go there either in the forenoon of a day when the northwest wind blows fresh and strong, and brings down fleecy clouds through a deep blue sky all the way from Greylock and Hoosac, or else to wait until the shadows are slanting down the hill with the westering of the sun on a clear afternoon. The bracing northwester promises the widest outlook, the most abundant play of light and shade, the liveliest music among the leaves. But the afternoon hour is enlivened by the sweet symphony of the woodland birds, and the dreamy haze on East Mountain and Tom Ball is as potent as a magic spell to him who is susceptible to the hypnotism of nature.

One always feels a certain diffidence in introducing others to his favourite books, or scenes, or friends. Taste is a most uncertain element of character, and may not be too confidently counted on. I feel the familiar misgivings as I find myself climbing the rails and setting foot with strangers on the lovely hillside. Perhaps others may not see it with my eyes, or feel how subtle and delicate a charm it has. But if they find it a tedious walk they can go back or go on farther. The village is scarcely half a mile away; and over the ridge lie the mountains.

But before one can enter the charming territory he must pause at the moth-mullein whose silent challenge to the eye halts one at the border of the field. Just why this tall and soldierly weed has been set to patrol this edge of the pasture, I am sure it would be hard to tell. Only a small squad is on duty, deployed along the depression near the fence. But always I am arrested here as I was the first time I ever crossed these boundaries. Moth-mullein is one of nature's surprises. Like the harebell, balancing its ethereal beauty on the edge of a bare cliff, or the water-lily, extracting fragrance and purity from ooze and slime, this dainty blossom wins its delicate colours and exquisitely fragile texture from thin and unpromising soil, in this dry and exposed corner of the hillside. Folded up in the bud it is always the same flat little wad with hardly a suggestion of possibilities of grace or delicate structure. And it is a witness to the regularity and constancy of nature that in that folded bud the outside fold turns down from the top, as regularly as the gummed flap of an envelope. I can never pass this group without plucking one for a souvenir ; long before it is landed in the vase at home its open petals have withered or fallen off. But once in the nourishing water its buds unfold in slow daily succession, and it holds its fairy-like beauty for a week.

Once past the mulleins, one comes to the peculiar glory of this hill pasture, the scattered bushes of the shrubby cinquefoil drawn up in a careless "open order" over the slope. In some moods one is inclined to criticise this leisurely bush because it takes so much time for its blossoming. The limpid yellow of a few blossoms dots each one of its bunches of green leaves ; but there is always lacking the splendid blaze of flame which would illumine the field if only they would all burst forth at once. Still, nature's method has its own charm and advantage. Looking over the hillside as it lies one can easily imagine that on the last clear night the midnight stars photographed themselves in colour upon this green and that the dome of the hill reproduces the concave of the heavens.

For pathways through this open jungle of shrubby cinquefoil one has choice unlimited. But chiefly the feet are drawn in one of three directions. There is a tempting prospect up the hill to the right, where a lovely elm, with trunk all thick and green with foliage, rises against a background of dense woods. In the shady afternoon those thickets are so cool, so suggestive of ferns and vines, creeping things and things that are fragrant with woody odours, that one inclines that way.

Straight on, the clear, clean ridge cutting the sky prophesies a good outlook across to the mountains. To the left lie some youthful pines, hobnobbing with young birches, under whose branches, just leafy enough to cast a shade, one may stretch upon the turf and hear the idle gossip of the winds. Hither I love best to stray. I have a natural affinity for pines, and there is no better accompaniment for an hour's day-dream over the landscape than the voice of these half-human trees. With that sound in one's ears it were possible to paint almost any fancy-picture of the land which lies behind those eastern hills.

I wonder if other people happen to share my preference for particular points of the compass? From my childhood I have loved the east better than any other point, and the horizon where the sun begins his day has a distinct claim over all the rest of the circle. Perhaps it is because, when a little boy, I used to gaze over the eastern hills from the garret windows of my father's house, and long for a glimpse of the fascinating ocean, which I knew was forty miles away beyond those ridges. I never confided this prejudice to anyone. So I cannot say whether it is a common one. But perhaps Mr. Francis Galton, who has discovered so many curious freaks of the human intellect,—hatred of certain colours, passionate love for odd numbers or even, association of colour with days of the week,—might tell us that this is no uncommon trick of the sentiments. At all events, I am glad that this favourite outlook is toward the east. I can lounge here in body, and thought will fly, far swifter than the winds, across these Berkshire hills and dales, over the middle counties of the old Bay State, down to the waves which wash the Gurnet and the sandy bluffs of Cape Cod. Then, with closed eyes, and ears full of the sea-sounds in the pines, I can almost believe that my body, too, has been translated, and that the old Atlantic is breaking in a lazy surge just down by the stone-wall.

But the sun moves around and drives one from the shady covert and puts an end to these dreams by daylight. Yet others follow hard after. Even as one lies here with cheek close to the earth, there steals upon the sense a fragrance, pungent, aromatic, subtle as some rare perfume, and elusive as the flight of the firefly. It calls to memory the interior of some country homestead, and conjures up the cupboard where the "simples" are kept, and clean cool chambers with beds whose linen gives out this same sweet exhalation. One has not far to look for the fragrant everlasting whose woolly blossoms yield this pleasant breath, dear to every country boy and girl, but dearer still to him who hides a bunch of it in the desk drawer at the city office, a swift reminder in the busy hours of the far-off hillside under the summer sun. Gather a handful of this grateful yield of the hill pasture, and stroll a few rods farther, for in yonder copse is reserved a pleasant surprise.

Pass the chevaux-de-frise of the birches, and work your way down the steep bank for a few rods, and you shall find yourself in the midst of a growth which florists would give much to discover, and which no lover of the woods would ever disclose. It is a splendid patch of maiden-hair ferns, covering many a square rod of the thicket, multiplying and luxuriating in the leaf-enriched soil. Hidden away from the eyes of the careless and the vandal, growing and fruiting and growing again for many a year, these delicate and graceful fronds have possessed themselves of this spot. It is their homestead. Inherited from generation to generation, the copse is the ancestral home of this delightful family, where they still rear their bright stems and spread their dainty pinules unharmed of men. What keener pleasure awaits the wood stroller than such a corner as this, redolent of the life of the finest flower of the shadows, fragrant with the gathered tradition of this family of ferns, whose very presence here attests how rare have been human visits, how largely this place is secured to the dryads and their mysteries? Is it not worth a walk over the hill-pasture to stumble upon such a woody corner as this?

7. THE CIRCUMVENTION OF GREYLOCK.

The very grandeur of mountains lies in their height, mass, strength, and sky lines, and none of these is seen so well from the peak as from the valley.

JOHN C. VAN DYKE.

THE CIRCUMVENTION OF GREYLOCK.

SOME folks think that the only respectful way to treat a mountain is to climb it. They feel in duty bound to scale every accessible height. They regard the failure to do this as a disregard of the challenge which the mountain tacitly implies to every able-bodied person in the bulk it rears before his eyes. The Alpine Club has done much to foster this idea in Europe. The Appalachian Mountain Club has cultivated it in this country. But the theory ignores one great fact in history. The greatest results in human progress have been achieved, not by climbing mountains but by going around them. Empires progress, civilisation advances, not by surmounting the great ranges but by circumventing them. The world moves, not by way of the peaks but by way of the passes.

This is not the reason why the two of us went around Greylock. We really were not caring much about civilisation anyway. We were out for a good run on our wheels, and we had been told that there was plenty to interest in the north end of the county. I had skirmished in advance in an earlier trip; and now The Lady was going with me. We took the way around, and the longest way around, too, in order to get views. One of the chief uses of a good mountain is to afford views. Mountains have a distinct utility as objects to be looked at; and it may be fairly questioned which of the two better knows his mountain, the one who has crawled and camped all over it, or he who has travelled and loitered all around it. Certainly the climbing brings only a half knowledge. A mountain needs to be seen in perspective.

The wind was cool from the north-west as we wheeled past the Maplewood and congratulated ourselves on the mingled self-denial and frugality which had induced us to be contented with a railroad lunchroom instead of the homelike dining-room of the fine old hostelry. But we saved an hour by lunching instead of waiting for dinner, and were well up to Pontoosuc, with its sentinel pines, by high noon. There was a fine blue ripple on the lake, and miniature whitecaps flashed in the sun; and away to the north the rounded shoulder of Greylock rose like one of the bounds in the *circus maximus* in which we were to do our chariot race.

The fine air was like wine in our veins, and as our machines worked free and the whirr of the chains and the click of the bearing-balls rose to our ears, we began to feel all the intoxication of spirit which comes to the well-mounted wheelman when he knows that all is well overhead and underwheel, and a good road to the fore.

It was an ideal track we took that afternoon,—one

Mount Greylock, Adams.
(From the "Back Road," Adams.)
"*A mountain needs to be seen in perspective.*"

of Berkshire's best ; and that is saying much. The little side-paths, beaten hard with recent showers, were smooth as roadways in a park. The grade through Lanesboro and part of New Ashford, though rising slightly, was easy always, and often varied by little slopes and plunges, long enough for coasting, with its exhilarating relief. Quiet homesteads and comfortable farmhouses gave a sense of companionship, though few people appeared anywhere about them. The rugged foothills of the Greylock range began to look quite near and neighbourly, but set us to wondering if our way were not tending toward hard climbing, when the direction to take the "second turn to the left and first to the right" swung us away from these laborious-looking regions. The road did what all sensible roads do ; it slid around the shoulders of the hills and glanced off from their precipitousness till it found a brook. Then it ran where the brook guided, and took all the easy grades, the cool ravines, the broad meadows.

And we who followed the road were refreshed continually by the gurgle and the ripple and the lapping and the dashing of the stream, hidden most of the time, sometimes far below us, sometimes above, now on one side, then on the other, till the little hamlet of New Ashford suddenly materialised out of the woods. We had a pang as we passed by this ghost of New England's old village life. It was such a bygone village. It was so lonesome in its grey and weather-beaten isolation. It was so forsaken of the industries and so bereft of all social stir, that I felt like hurrying away from it, as from a scene of pain or privation. But we had to stop. We were forced to pay the local scenery the highest tribute a confirmed bicycler can offer, by dismounting to look down into the deep, cool, moist ravine through which Green River drops to lower levels. Under rich hemlock shades, through a moss-grown gorge, the waters sluiced their way swiftly to the valleys below.

But we could not watch them long. The Lady would rather coast than eat, much more than look at the finest scenery. She was up and away long before I was ready to leave this glorious gorge ; and I perforce must follow, albeit one or two curves behind her, but assured of her safety by the tinkle of her bell as she warned imaginary wayfarers of her coming. And it was only by imperative language, and almost physical force, that I could induce her to dismount for the next hour.

For now we had struck the down grade to South Williamstown ; and for seven good miles we were sliding down-hill, much of the time with our toes on the coasters, in a glorious swift flight. The mountains closed in upon us in wooded gloom, on either hand. Greylock and his attendants were wholly hidden behind a half-dozen intervening ranges of hills. Wild woods hedged our way. The river kept us company. Here and there a farmhouse broke the solitude, and the voices of haymakers and the clatter of mowing-machines came from far-off fields. And over all was the deep blue of the sky, and drifting fleets of cumulus, wafted by the light breeze which fanned our beaded brows. Oh, it was fine! We never expect to transcend the sensations of that run, till perhaps we get our wings and learn to fly. Once we dismounted to sit five minutes beside a tributary brook, and pass bantering speech with an inquisitive country lad. Once we stopped to tender sympathy and aid for a collapsed tire to a brother wheelman— the most demoralised and hopeless tire I ever saw.

Then the ravine broadened into an intervale, and the intervale into gracious meadows, and before we could well realise where we were, the noble elms of old Williamstown were arching over our heads, and Greylock was behind us, and full in our sight, eastward toward North Adams, rose the dignified forms of its companion hills, Mount Williams, Mount Fitch, and Mount Prospect. A college town without its population of young men is a lonesome place; yet we felt the absence of the wise and distinguished youth and their professors less than we might have done but for a revelation of the presence of a sacred classic character whose proximity was accidently betrayed to us. For as we explored the vicinity of the Campus, a sunburnt boy bearing a tennis racquet appeared on a sidewalk, shouting across the street, "Say! Where's Homer?" And when a voice replied, "He can't come out yet," we felt that we had met with a keen disappointment. Yet we went the happier out of that town from a feeling that it was harbouring the "blind old man of Scio's rocky isle." It was a fit haven for old Homer, and we were glad to know that the rumours of his demise were unfounded.

We would have stayed longer, on the chance of seeing the old fellow. But there were still six good miles between us and our goal at North Adams, and we pushed on to the charming home whose hospitable roof received us long before supper-time, the richer by an afternoon of untempered outdoor joys.

I pass the Sunday which followed, when the wheelman was metamorphosed into a parson, and preached under the shadow of Greylock. By night the clouds returned and the rain fell, and Monday morning was ushered in with muggy airs and a blurred and uncertain sky. But the itinerary had to be kept somehow, and the circumvented mountain photographed. We planned to wheel as far as Adams and pick up a train for Pittsfield there. But, alas! The soft and sticky road and the softer and stickier air cut down all hopes of a speedy run, and long before we had accomplished half the distance by the "back road," the warning whistle in the valley below told us of our defeat. But we consoled ourselves by one or two tries with the camera at Greylock's round shoulder, as it rose, not unlike our beloved Dome in southern Berkshire, the splendid background to a broad and lovely valley.

Then came more mud in the roadbed, and more language on the lips of the riders, which rose—the language, not the mud—into a torrent of pious expletives, as we picked our cautious way down a breakneck road, crooked, gullied, rocky, slippery, edged by a threatening guard-rail and a plunging brook. When we emerged, undamaged and undismayed, into the level streets of Adams and quaffed our mild tipple at the soda-fountain, we heard of two riders who had come to grief and fractures on that shocking thoroughfare, and felt with great satisfaction that we had really performed a feat.

But there were no more adventures, no thrills and no spills in the rest of our humdrum push on to Cheshire. There were one or two glimpses of Greylock above the nearer hills, and some sweet, wholesome stretches of woods. But the air was lifeless, and our clothing was drenched, and we were a little impatient of the stern prose of our wheeling. The machines were no longer wings, but weights, and we could not spin, but only plod. So when we were up with the station at Cheshire we felt that we could honourably abandon a heavy road and hurry on to Pittsfield. We had accomplished our purpose. We had circumvented Greylock. We bade the old fellow a loving farewell as we dropped him behind, which was in our thoughts a half-promise to return,—an *au revoir* to his excellency.

8. BERKSHIRE GLIMPSES.

What floods on floods of beauty steep the earth and gladden it in the first hours of the day's decline !
>Thomas Wentworth Higginson.

BERKSHIRE GLIMPSES.

IT was a day which called us out of doors and up the heights. A transparent atmosphere brought far places near and tempted the eye to the most remote horizons. When one can include the far-away in his range of sight it is his duty to do it. There are times enough when clouds, mists, and darkness confine the eye and it must perforce content itself with what is near at hand. But when the skies clear, and the sun rises, and the mists dissolve, the time has come to look abroad and re-establish relations with the remote but still real world. It hardly needs to be added, brethren, that this truth has a spiritual application.

Two miles or so from our home there is a steep little cliff which terminates a range of considerable mountains, known even upon the dignified maps of the State survey as "Jug End." Its summit is a bare ledge, facing squarely to the north, from which the tramper's instinct assured us there must be a fine outlook. Local opinion discouraged the attempt to climb the ridge, and prophesied that only fatigue and failure would come of the expedition; but that is what local judgment always does. It is rarely that the "native" looks with much hope and good cheer upon the aspirations of the summer-boarder toward mountain-tops and rugged heights. It is generally his function to reverse the office of Stockton's official and become the discourager of precipitancy. Only when much experience has taught him that good muscle and wind are often to be found under an outing shirt, does he concede the probability that the "city chap" will scale any peak he attempts. So we waived the disparagements which greeted our announcement of determination and turned our steps toward Jug End.

The walk thither was entrancing. It lay through the levels and the uplands rich in July verdure, through grasses ripe for the mower, and tangles of wild-flowers crowding the wayside ditches with their aggressive ranks, lining up and touching elbows for their annual charge upon the farmer's hardly conquered domain. The wild-flower is the barbarian of the soil. It holds the same menace over the civilisation of the fields that the Persian held over Greece, the Gaul over Rome, the Kelt over Britain, the Indian over America. Let the civilising and cultivating hand be withheld for but a season, and the barbarian weed is over the wall, under the fence, throwing a skirmish-line up the hill, and deploying down the valley. These wild things have a system of warfare all their own. They are invaders by nature. They have been trained for nobody knows how many million years in the art of beating any form of life less energetic than themselves. Their tactics are those of the savage; they overcome by craft or else by sheer force of numbers. And any farmer will tell you whether they are fighters or not. Look how the common daisy has forced its way stubbornly into the fields of this continent. One has only to look afield to see how in the farmer's territory as well as in that of the sociologist eternal industry is the price of civilisation.

I do not say that such reflections as these added their gravitation to our steps as we clambered toward the woods which tangle the ledges of Jug End; but the text was there, whatever became of the sermon. Nor was it long before we encountered a new application of it in the dense thicket which covered the steep and rocky mountainside. Here, too, nature was getting even with her despoiler and sending in legions of sturdy warriors to recapture the lands she lost in the extermination of the hemlock and the pine. Scrub-oak, maple, moosewood, hobble-bush, and blueberry,—they were all here, the hardy Cossacks of the forest and the hillside, throwing themselves recklessly into the fight against the usurper, man, and holding possession of these strategic points until the reserves shall come up, the slow-moving evergreens, the sylvan infantry. And there they will hold their ground if need be for ten thousand years.

But it did not take a long nor a particularly hard clamber to pierce this natural abatis and reach the crest of the ridge which juts out into the level meadows of Egremont. Once there the wide prospect, north, east, and south, was a very dream of beauty Above, the creamy clouds were sailing down a clear blue sky before the fresh wind of the north, and below, the shadows were chasing one another over the farms and villages of fair Berkshire. Such a landscape one sees this side the sea only in New England ; and only in Old England on the other. It is the landscape made possible by generations of patient toil, the result of many ploughings and reapings, much seedtime and harvest, smoothing the rough earth into comely beauty, trimming down the asperities of the forest, piling the meadow stones into those ramparts which are the sign of the industry by which man has made the unremunerative earth into prosperity. It is such scenery as man has had a share in creating. It is the token of his victory over the lower world.

Here are the clustered houses of half a dozen villages and towns, each with its church-spire, like a lifted finger in testimony of the ideals which have inspired this toil and led up to this visible result. Here are the outlying farmhouses, nested each in its clump of trees, whence so much energy, intelligence, and moral impulse have flowed into the life of Massachusetts and of the nation. Here, too, the yellow rye-fields, the green acres of the ripening oats, the meadows from which the busy clatter of the mowing-machine rises to the ear, all tell a story of prosperity still unchecked and country life still bestowing industrious and thrifty effort on the land.

Jug End, Egremont.
(Looking toward Norfolk Hills.)
"*Such a landscape one sees only in New England.*"

But there is a background to this fair scene which the eye of the lover of the hills seeks with joy and sweet content. Westward from our outlook the mountainside falls away into a deep glen, whose sides have been cleared well up the slopes and converted into comfortable meadows. Beyond the valley, forming the western wall of the ravine, rise a thousand feet of mountainside, thickly clad with chestnut and maple, whose brilliant greens have grown soft with the afternoon haze since we sat on the summit, and fill the fancy with hints of rest, of perfect quiet, of serene repose within their leafy depths.

The breeze lulls for a moment ; the far sounds from the farms come to our ears softened and sweet. But best and dearest of all sounds, across the glen, from out those woody coverts, there floats the tender, liquid trill of the thrush. It is the harbinger of the evening, the first notice the birds serve that the day is waning, and that the shadows are gathering in the forests on the eastern slopes. There is no other woodland note like this. It is perpetual music. It touches the emotions like profoundest poetry. It calls on the religious nature and stirs the deepest soul to joyous praise. There is no bird, among the many which have found their way into song, in other lands or other times, whose note deserves so much of poet and lover of nature as the wood-thrush. The very spirit of the forest thrills in this vesper-song. It is the trembling note of solitude, rich with the emotions born of silence and of shadow, rising like the sighing of the evergreens, to fill the soul at once with joy over its sweetness, and with sadness because that sweetness must be so evanescent. When one has heard the song of the thrush there is no richer draught of joy in store for him in any sound of the woods. There is nothing to surpass it, save the ineffable ecstasy of the silence which reigns in their deepest shades.

There can be no more fitting climax to the pleasures of a day in the fields than to hear the thrush. It is the Angelus of nature. Let us go silently home.

9. A MAY-DAY ON MONUMENT.

    Yet "God be praised" the Pilgrim said
        Who saw the blossoms peer
  Above the brown leaves, dry and dead,
      "Behold our Mayflower here!"
                          JOHN G. WHITTIER.

A MAY-DAY ON MONUMENT.

I HAVE always had great sympathy with the boys who march at the head of processions—that inevitable company whose presence is as certain as it is incongruous, and whose chief joy appears to be in the fact that for them the procession is always an anticipation. It is yet to come to pass. It is still future. These youngsters are epicures in sensations; and they take keen delight in maintaining that which is to the sidewalk spectator a show, first passing and then past, as a pleasure to be tasted, a sweet morsel still rolled under the tongue.

Now if I had the means and the time, I should every year in this same fashion run ahead of the vernal advance, the procession of leaves and blossoms and birds and butterflies, as it moves northward from the Carolinas to the Canadas. There is such an exquisite pleasure in watching the burst of life, the outbreak of colour and fragrance, the clothing of field and forest with verdure, that one would be glad to prolong the sensation. In these days it would be an easy matter to keep just ahead of summer for a good two months. And then one might halt on the banks of the St. Lawrence and let the pageant pass by ; for when it has gone as far north as that, the line of march is nearly done.

I have a friend who once did the next best thing to this,—some will think it a better thing. She began eating strawberries in Texas, I know not how early in the spring. She ate through the season there, and then, coming north, found it just beginning in New York. She lingered there, still eating her fill of strawberries till they began to wane ; and then she moved on to Maine, to find the native and the wild berries just in their prime. It seems almost as if a way had thus been discovered to eat your strawberries, and have them too !

I have been making a little essay in this direction myself. After seeing the buds unfold and the violets bloom and the forsythia pour its golden rain in the city's parks, I took a short-cut by rail to Berkshire, and there intercepted the head of the procession, and repeated the delights of seeing the column of spring's splendours passing in review. And the ostensible object of this frivolous excursion was a little breathing-space after a breathless winter. Its only visible fruit is a bunch of arbutus and trilliums. I cannot expect any great sympathy when I say that to me they are more to be desired than trout, yea, than much heavy trout. But such is the fact. They are blooming in the parlour now, and for a week perhaps will give their silent reminder of the woods to every incomer ; no fish could be exhibited as long as that after he came out of the water ; he would not look well, and he would not be in good odour. Nor do I believe that any trout ever gave more pleasure to the hunter. Did you ever hunt the trillium? Did you ever seek the arbutus in its forest haunts?

I found my trilliums in a wood, less than a quarter of a mile from the road. You go in by a path which begins at the corner of the wood, and almost immediately find yourself under the high spreading boughs of a clump of hemlocks, tall, sombre, and mysterious. They have a whisper like the pine; and like that tree they also, in the struggle for light and sunshine, lift the most of their branches skyward, and are almost bare of limbs lower down on their trunks. Beyond them is a sluggish stream where the marsh marigold is blazing in great spots of yellow, and where by-and-by the fringed orchis will come, and the cardinal-flower. White and yellow violets dot the grassy wood-road which forms the path, and the ferns are beginning to uncurl which in July and August will wave their graceful fronds in the lightest zephyrs. Just now the ferns are interesting, but not graceful. They are long and lank and fuzzy, and remind one of nothing so much as young colts, with their thin legs and frowzy coats.

But the trillium patch is a little farther on. It is time to look for it now, on either side of the way, deep in the tangled thicket or close to the grassy sward. There is one, thrusting its long, lily-like stem up through the matted leaves, its bud rolled up into a sharp point, so as to pierce the tangle more readily. When it is six or eight inches up it opens its three dark-green leaves, and a little later unfolds its lovely blossom. Three pure white pointed petals, alternating with three equally acute sepals, raised an inch or more above three green leaves and carrying twice three stamens and a three-parted pistil, that is the structure of the trillium. All good things go by threes with this lovely blossom; and there was reason in the inquiry of the small boy to whom I was showing them on the way home, and who asked, "Did you call them 'triplets'?" With the accurate mathematics of plant-life, they certainly produce all things in triplets.

But I ought to have said that the plant I am after is not the deep purple variety, the "wake-robin" of the books and the country people, but the "painted" trillium, so-called from the delicate pencillings of crimson about the base of its snow-white petals. If it were only to become a fad, like orchids, or chrysanthemums, or the maiden-hair fern, it would richly repay the enthusiasm of amateurs and the zeal of cultivators. It would grace the button-holes of elegant young men, and adorn the vases of charming women. But it is better far, to all true nature-lovers, just as it is, the child of the early spring, the ornament of damp and boggy woods, the one spot of colour and beauty where all else is sombre and plain. One plucks it to a running accompaniment of bird-notes. As I vibrate from side to side, reaching into the underbrush, making detours round swampy spots,

I hear the cry of the hawk, and see his ominous shadow darkening the leaves. The note of a thrush thrills the still air ; and as I stretch my arm toward a splendid blossom, I start back at the tremendous drum of a partridge on which I have nearly placed my hand, and which goes thundering away into the deeper thicket.

For the arbutus hunt, a special day was set apart, and a special company bidden, choice spirits who love the woods and are victims of what Kipling calls "The old spring-fret." We fixed on May-day, and planned to strike the trail of the mayflower on Monument Mountain. The farmer's big "three-seater" bore us across country and up the fine road from Barrington toward Stockbridge. It was one of those hot, unseasonable days which anticipate in May the coming summer, and the horses panted and sweat under the baking sun.

The green things of the soil,—turf, leaf, and fern,—seemed to take a great leap forward as the hot wave rolled over them, and we could almost see things grow all around us. The air was full of the aroma of dank earth, of delicate blossoms, of soaking moss, of saturated mould. When we turned aside from the road and took the wood-path up the mountain, the sun, pouring down through the forest trees as yet showing hardly more than buds, seemed to create a perfect ferment of growing and unfolding germs in every rod of soil.

But now began the joyous hunt for the arbutus.

From my earliest days I have had an intense affection for this flower. I have been passionately fond of its delicate blossoms. I have craved its faint but satisfying perfume. I have revelled in its dainty colouring. I have loved its modest, shrinking habit, its vain attempts to nestle under the dead leaves, and hide itself in the shelter of its own foliage.

Perhaps my affection is inherited and rightfully belongs in the traits of a descendant of the Pilgrims; there may be an ancestral pulse in the thrill which comes at the sight of it in spring. Perhaps it has gathered to itself the associations of many a boyhood hunt in the woods and on the rocky hills of the dear Bay State. But whatever the source of the sentiment, it renews itself every spring with something of the persistence of a religious devotion. I am almost sorry that I compared the arbutus, a moment ago, to a fish, even though it were a beautiful fish. There is something carnal about a fish; and there is nothing, absolutely nothing, that is not dainty, subtle, ethereal, in the make-up of this flower.

How welcome, then, this fresh prowl among the dead leaves, this search for the glistening green ovals of the arbutus' leaves, the faint whiffs of perfume as we unearthed the little blossoms. The date was a trifle late to ensure our finding the flowers at their best. We had reckoned a little too confidently on the high ground and the woods as likely to retard their blossoming. But there were enough of them, and we worked our way up the slopes toward the south-

Monument Mountain, Stockbridge.
(Looking across Flooded Fields, Great Barrington.)
"*It dominates the meadows of Barrington and Stockbridge.*"

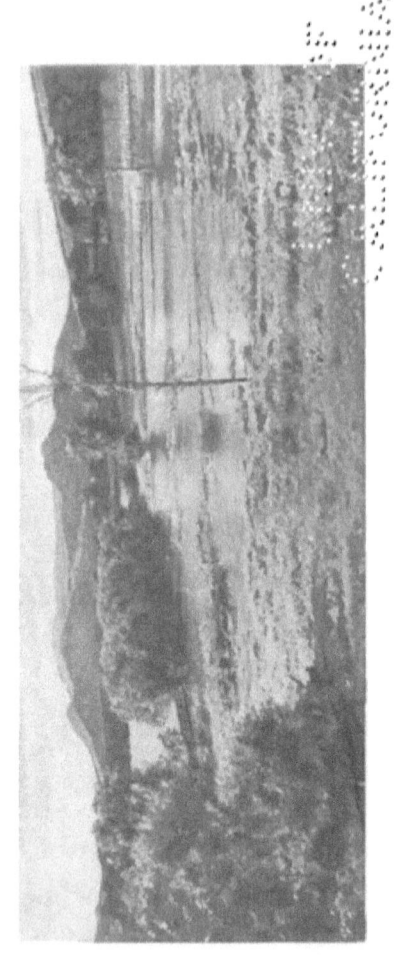

ern peak of Monument, culling delicate blossoms all the way.

It was high noon when we sat down on the rocks which overlook the steep precipices of the eastern side of the mountain, and spread our frugal lunch. Almost everybody who knows Berkshire with any thoroughness has studied the landscape that drew our eyes away from our rocky table to the hazy horizons. And nearly everyone has seen it through the eyes of the Berkshire poet, who tells of it in words that make other description superfluous:

> "It is a fearful thing
> To stand upon the beetling verge and see
> Where storm and lightning, from that huge grey wall,
> Have tumbled down vast blocks, and at the base
> Dashed them in fragments, and to lay thine ear
> Over the dizzy depth, and hear the sound
> Of winds, that struggle from the woods below,
> Come up like ocean murmurs. But the scene
> Is lovely round; a beautiful river there
> Wanders amid the fresh and fertile meads,
> The paradise he made unto himself,
> Mining the soil for ages. On each side
> The fields swell upward to the hills; beyond,
> Above the hills, in the blue distance rise
> The mountain-columns with which earth props heaven."

All this we saw in the very prime of the spring, in the haze of a May-day which was premature July, with the brown acres of fresh-turned soil telling of the seed-time just passing by. It was our last chance to see the procession of budding things march past. In a few hours we were whirling homeward. But we bore with us much trillium and arbutus, trophies of the woodland and memorials of the passing spring. And in those brick caves which city dwellers call houses, they bloomed for many days, as reminders of what the brown woods yield to the first caresses of the spring.

10. AMONG THE CLOUDS.

I am the daughter of Earth and Water,
   And the nursling of the Sky :
I pass through the pores of the ocean and shores ;
   I change, but I cannot die.
For after the rain, when, with never a stain,
   The pavilion of heaven is bare,
And the wind and sunbeams, with their convex gleams,
   Build up the blue dome of air,
I silently laugh at my own cenotaph,
   And out of the caverns of rain,
Like a child from the womb, like a ghost from the tomb,
   I arise and unbuild it again.

                                  PERCY BYSSHE SHELLEY.

AMONG THE CLOUDS.

ONE of the prime requisites of an ideal summer home is a wide horizon circle, and an unobstructed view of the firmament which domes it. For nobody lives out of doors, in any true and large sense, unless he is constantly in touch with the sky and the clouds which so often fill it. The wise and affectionate devotee of nature will therefore pay cheerfully a good round tariff for a fine sky-prospect, and bear much weariness or appetite at the table for the sake of a good diet of clouds.

For one, I have always had a passion for clouds, and, without being a person of glaring impracticability, I have passed much of my life among them. So that it has come to be an indispensable feature of every well-spent day to consult, and study, and hobnob with the clouds. It is a function which I regard as partly a duty and partly a privilege. For it is a man's duty to know something of what is going on in the world around him, the signs of the times and the *status quo*; and there are no truer or more reliable guides and interpreters in the external world than the clouds. Nor can any higher privilege be given one than to be furnished with the key to even a fraction of what these travellers in the upper world have to say as they go their various ways. So it would be a sad thing to be hindered in one's duty and balked of one's privileges all the summer, by an unfortunate choice of location which should circumscribe the horizon and limit the sky-view.

It is not too much to say that, for all sensitive natures, keyed to a nice and intimate association with the external world, the day takes its complexion from the sky. There is an acknowledged connection between the face of the firmament and the aspects of mountain and sea, ay, of the whole landscape. When the clouds gather, the sea looks black and the mountain frowns; when they scatter, the waves seem to dance and to smile and the mountain glows with new attractions.

There is an association quite as close between the moods of the human spirit and the face of the sky. The coarsest natures feel it in relation to the sharper contrasts in this upper zone. The veriest lout will scare a little when the thunder-squall darkens the noon. Even kitchen scullions brighten up a bit when the long storm breaks and the sun shines again. But this is only the rudimentary sense. Refined and developed, it becomes alive to the finest shades of difference in the light, the draping of the clouds, the transparency of the atmosphere. I have known a man to be homesick as long as the haze and smoke hung low over the mountains and hilltops, and recover all his spirits as soon as the north wind brushed the air clear and bright. There are days when the aspect of the clouds fills me with nervous anxiety ; and there are others when every look into the sky yields a revenue of courage and hope. I do not undertake to account for this fact. I suppose it comes through the associations of the mind and grows out of the memories of other days and skies, which have come to form a sort of sub-consciousness.

I can believe those people who insist that the weather makes no sort of difference to them, and that one sky is just as good as another ; but I cannot enter in the least into their feelings nor understand their indifference. They seem to me defective, in some such way as those are who cannot distinguish between "Yankee Doodle" and "Old Hundred," and would as soon hear a hand-organ as Seidl's Orchestra. They usually boast of their lack of sensitiveness as a sort of superiority. But to be cloud-blind seems as bad as to be colour-blind ; and not to be moved by the aspects of the sky is as unfortunate as to know no difference between the smile and the frown on the face of a human being.

To-day, for instance, we have had a fine succession of cloud-effects, whose changes have been like the moods of a friend. The morning opened with lowering clouds, low, dark, in hurried movement, pouring heavy showers in copious bursts with each fresh squall. It was one of those rains which had no threat of thunderbolt or of tornado, and all the cloud aspects were friendly and reassuring. The rain drew a steel-grey curtain between the eye and the landscape, which added a distinct charm to the whole circle of scenery. All the face of the sky seemed to say to the beleaguered human beings in the house: "Stay at home and be happy; read, chat, sing, mend stockings, play games, while the earth soaks in refreshment and nutrition from the clouds." There is a sense of the comfort of home which one might call storm-begotten. I fancy most people learn to relish it thoroughly. It puts a premium upon all the attractions of the sheltering roof and the cheerful fireside. Yet the clouds have no menace of harm to the earth. They only seem to deprecate going out of doors. They encourage home-staying.

But soon after breakfast, almost before the discussions and prophecies about the day's probabilities were out of the way, there came great rifts in the clouds, through which the deep blue of a clear upper stratum gleamed brightly and hopefully. The rents grew larger, the sunlight came in more frequent bursts, the tattered edges of the storm-cloud hung darkly in the east, while westward the blue was bright and cheery with the promise of clear weather. I could not help noting, as I studied the retreating wrack, how much a certain blue patch in the cloud-bank resembled the angry glow of an inflammation. It was like a bad symptom, in this instance just beginning to disappear.

The early afternoon brought a new phase of cloud-

Clouds on The Dome, Sheffield.
(Looking south from Egremont.)
"*Fraying out into mere ravellings of vapour.*"

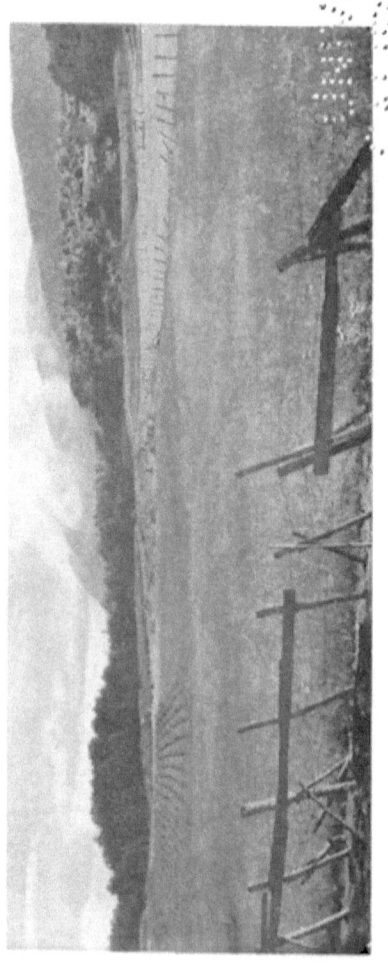

scenery. The soft cumulus clouds which had been hardly more than shreds and patches all the morning massed and moulded themselves into great bastions and bulwarks of brassy-white, deepening into a copper-colour as the hours went by. Such a sign is seldom unfulfilled, and it was no surprise to hear before long the boom and rumble of distant thunder. The squall gathered squarely in the west and advanced directly over our territory. It was one of those sharp, decided, business-like squalls, which drops a few bolts from its advancing edge, and one or two in retreat, and then is gone; like the light-batteries in the rear-guard of a retreating army, whose vigorous fire holds the enemy in check and gives time for their own swift withdrawal.

Of course the sunset of such a day was full of quiet beauty. There were no such gorgeous hues, such lavishness of colour, crimsons, purples, and golds, as herald the approaching gale. But the west glowed with warm colour which tinted the few gathered clouds and rimmed them with gold; and the glory lingered, and slowly faded, long after the evening star had flamed out, and the moon had risen in the east. Who could look upon those mild hues and note the languid shifting of the scant clouds without a rising sense of security for the night, for the morrow, ay, and for all nights and morrows? For there is no feeling of permanence about a storm; but good weather rouses a responsive conviction in the soul that it represents something normal in the creation, an estate of ultimate fair weather, unruffled by the ruin and havoc of storms, under those "new heavens" which shall shine when "the former things have passed away."

11. THE SOCIAL FLOWERS.

A flower is organised for coöperation. It is not a distinct entity but a commune, a most complex social system.

But in still more ingenious species the partners to a floral advertisement sacrifice their separate stems and cluster together on a common head.

Henry Drummond.

THE SOCIAL FLOWERS.

"THERE," said the preacher, as he held up a handful of flowers before the loungers on the piazza, "is a prophecy of the Chicago strikes and of the Chicago Exposition. That bunch of wild blossoms out of our pasture foretells New York and Boston and London. They blazed the path for the founders of the American Union. It is not too much to say that they foreshadow the kingdom of heaven."

Nobody seemed to make out what the dominie was driving at, and so nobody made any remarks or comment upon what he had said. Nobody is expected to listen very carefully to what dominies say on their vacations; and it was the general impression that this sententious speech was nothing but a poor paradox, veiling some attenuated joke or other, such as the clergy off duty are much given to. So there was an inquiring silence while the company waited for the explanation which they knew was sure to follow; and when the stroller had seated himself and displayed the spoils of his afternoon's campaign, he expounded his paradox. In the words of the newspaper reporter, "among other things, he said":

"It is a random garnering from the roadside and the field. Here is a spike of blue vervain and a spray of the bright agrimony. Here are daisies and chamomile, and an early aster or two. Those are the graceful umbels of the wild carrot, and these the clustering sweets of the milkweed. Red clover everybody knows, and 'butter-and-eggs,' as well. It is a miscellaneous handful and there are members of half a dozen different families in it. But they are all of them clustering flowers. In one way or another, each according to its own habit, these blossoms have gathered themselves into groups. They seem to have discerned that it is not good for a flower to be alone; and these various methods of clustering their brightness and colour are a clear demonstration of that instinct in their growth. The yellow bloom of the agrimony around its curving spike; the white sphere of the button-bush; the yellow-and-white disks of the daisies;—are so many various ways these flowers have taken of 'getting together.' They are the expression and the result of the vegetable tendency toward gregariousness.

"For it must not be supposed that these flowers have been arranged in such various and attractive forms simply to gratify the eye of man when he should come along with a power to perceive and enjoy their beauty. That old conceit of human minds perished long ago in the frosty atmosphere of the evolution theories. The flowers began to combine in groups, and associate in spikes, clusters, racemes,

A Berkshire Thicket, North Adams.
(At the Natural Bridge, North Adams.)
"*A random garnering from the roadside.*"

supposed that these flowers have been arranged in such various and attractive forms simply to please the eye of man when he should come upon the scene to perceive and enjoy their beauty. Hosts of human minds perished in the deadening atmosphere of the evolution theories. The flowers began to combine in strains and masses of spikes, clusters, racemes,

and umbels, a good many cycles before man was taken into the account. They had reasons far more cogent and pressing than merely to get ready to please the eye of a being of the future, whose advent was still a matter of several eons ahead.

"It was a practical question of getting a living, a real matter of life and death with these flowers, whether they should combine and live closer together, or scatter themselves in isolation along the stems on which they dwelt. Because every one of them depended, for the perpetuation of its kind, upon the visits of insects on their travels, who, carrying the pollen across from flower to flower, cross-fertilised the blossoms and so secured the continuance of the plant. That result was most certainly attained in the case of those stalks on which the blossoms were crowded most thickly. For there the hurrying bees and wasps and moths and smaller honey hunters, the agents of the earliest sugar trusts, could most easily take up the pollen from one blossom and dust it over the next. But more than this, these busy creatures were more certain to alight and try their luck for sweets on some spot where a cluster of flowers made the red or the white or the yellow more conspicuous than it could be in a single blossom. So the clustering flowers were more likely to be cross-fertilised than the solitary ones; and their offspring were more likely to survive than the children of the self-fertilised blossoms. Thus the tendency to gather in groups helped perpetuate the flowers in which that tendency occurred; and the flowers thus perpetuated helped to pass on the tendency to associate in groups. The great principle of association was established among the flowers. Individualism was at a disadvantage; collectivism began to work ahead in the race of life.

"Nor was this all. Some of the smaller flower-folk were still at a disadvantage. Only by crowding together in closest contact could they vie in attractiveness with their stronger, because brighter, rivals, and so advertise their presence to the travelling insects. But when some of them did thus crowd together into a dense head, the tendency to dense-headedness was started and continued; and something began to be added to that tendency. Some of the outer flowers were set apart for the especial duty and task of attracting attention, while to the inner group was given the work of secreting the honey which was the price paid for the services of the useful insects.

"That was the way in which the great family of the composite flowers came to be,—the family which includes the daisy and the sunflower, the golden-rod and the immortelle, the tansy and the chrysanthemum. Grant Allen calls these the most advanced, the most highly civilised of all the plants. They deserve the distinction. They have come to live in little communities, and they have reached the point of a division of labour. Every daisy by the roadside is a village of tiny flowers. In that village there are two sets of workingmen, the ray-flowers which serve to invite the attention of the passing collector of sweets, and the disk-flowers whose office is to furnish him with goods when he has been attracted. The wayside daisy embodies two of the most vital principles of progress, two of the laws of civilisation, two fundamentals of the family and the state, namely, association and division of labour, coöperation and specialisation. After one comes to know that fact, it is a trifle amusing to hear the modern movements toward socialism in human life set aside as passing fads. There is a pretty fair momentum to a tendency which goes back to the foundation of the daisy communities. We shall find it hard to check a law so ancient. It is evidently a good deal involved in the nature of things.

"But that bunch of flowers carries by implication another hint of what we have deemed a very modern notion, a bright device of humanity to increase human efficiency, enrich human life, and foster human unity. Society in our modern thought implies intercommunication, transportation, travel. These are essentials of human life in its coöperative stage. They are all provided for here in the life of these flowers. The peripatetic insect, the restless bee, the desultory butterfly constitute the vast transportation system of this floral world. The business which they do is immense. Their traffic is beyond computation. Stop the wheels of their industry and the flowers would be annihilated.

"The freight lines of our great railways cut no figure whatever in comparison with the innumerable companies organised for the carriage of pollen-dust from flower to flower in the maintenance of plant life. The Black & Gold Despatch Company, the Wasps' Express Freight, the Moths' Night Line, are two or three of the corporations (unlimited) which transact the enormous business of moving a summer's crops of pollen. And when the season is over and the harvest is gathered it is the habit of the restless children of these flower communities to say good-by to their homes and to the village in which they have been born, and sally forth to found new homes and new villages for themselves. To do this they take their own private cars, run on the tracks of the great Air-Line System, and by way of the Great North-western, the East Wind Consolidated, or the South-western Central, fly to new lands to repeat the ventures of their progenitors. You may see them next fall, when the thistle floats on the breeze and the aster seeds start on their overland journeys. Who can fail to perceive in this vast machinery of the flowers, this coöperation with the bugs and the breezes, this intricate system of communication and transportation, a foreshadowing of all that the Vanderbilts and the Scotts and the Pullmans have done in human society?

"Now, to go back to the text, is not that handful of flowers a wonderful prophecy? Is it not a shadow of things to come? Does it not uncover the deep roots of law, the same back among the flowers in their unconscious life as in the vaster, higher life of man? Is it not a powerful confirmation of the faith of him who believes in the unity of nature and the solidarity of all things? Is not the principle of co-operation pretty well tested, if it is as old as the daisy-commune? If the tiny flowerets of the *Compositae* combine in their blossom-unions, shall not the workman in his trades-union? If the asters have proved that "in union there is strength," shall we not triumphantly look for a new vindication of the law in the life of the republic? If the ray-flowers of the sunflower have parted with some of their functions for the sake of the community, may we not see the foundation to the kingdom of heaven, the law of self-sacrifice, writ large in the nature of things?"

By the time he had reached this point, the dominie's congregation had diminished to one; the rest had stolen away to the croquet-field, or the tennis-court,—anywhere out of earshot. But she, the solitary little spinster, remained, and, as if she would show her keen interest and her aptitude for listening to such an extemporary treatise, asked in a casual sort of way, "Doctor, do you believe in evolution?"

12. THE BERKSHIRE RIVER.

Yet flowers as fair its slopes adorn
   As ever Yarrow knew,
Or, under rainy Irish skies,
   By Spenser's Mulla grew ;
And through the gaps of leaning trees
   Its mountain cradle shows,
The gold against the amethyst,
   The green against the rose.

JOHN G. WHITTIER.

THE BERKSHIRE RIVER.

THERE are some rather critical souls, who object to Berkshire because, they say, it lacks water. Having in mind, perhaps, the many lakes of the Adirondacks, or the endless chains of Northern Maine, these hydromaniacs complain of a lack of that diversity which is afforded by this feature in the landscape. Sometimes, too, they add a remark,—which betrays the real animus of their criticism,—that there is very little good trout-fishing in this region. I have my opinion of a man who only values a brook or a pond for what he can get out of it. And the demurrers of fishermen are always to be received with suspicion by real nature-lovers. The man who loves nature *and* fish is not open to objection; but the man who cares for only so much nature as he can reach with a trout-pole and line is not a competent judge of her charms. Such people overlook the picturesque Reservoir in Cheshire, with its stump-fringed shores and sedgy shallows; and Pontoosuc, the Coney Island of Pittsfield, and its neighbour, Onota; and Stockbridge Bowl, which even the proximity of civilisation and architecture cannot quite spoil; and Lake Buel, and Lake Garfield, and Long Pond, and

Winchell's Pond, and the Twin Lakes. But most of all do they neglect—and the oversight is unpardonable in a nature-lover—the stream which like a silver cord binds the scenes of Berkshire into one volume, the winding Housatonic. As long as this Berkshire river creeps through the meadows, and frets over the rocky shallows, and takes the shadows of overhanging cliffs, and plunges through mimic gorges and ravines, it will save the region it adorns from the charge of dulness and endear it to every open eye and ear. He who knows his Berkshire will never omit the praise of the Housatonic.

Great rivers do not lend themselves to personal affection; and they are too distributive in their effect to create much of any impression. One could hardly give a comprehensible answer if asked to describe the effect upon his imagination of the Mississippi, or the Amazon, or the Ganges, or the Nile. And as for loving any of these mighty streams, as one loves the Connecticut or the Charles, as Englishmen love the Thames or the Dee, the thing is absurd. One might as well try to be fond of the Rocky-Andes system, or to claim the great wheat-plains of the West as his favourite corner for a summer-resort. Little rivers are the only ones with which one can be on intimate terms, toward which one can be fond and friendly. They admit of comprehensive views. They can be grasped, in a certain unity of impression. A river like the Housatonic is just large enough to be significant, just small enough to challenge friendship.

In choosing the Housatonic as the typical Berkshire river, no disrespect is meant or shown to the other streams whose life is begun and partly passed in the county—Hoosic, Green, and Westfield in the north, and Farmington in the south. None of them shares the natural life or contributes to the scenery of this region for so long or to such a degree as the Housatonic. Its rise is in the northern hills of the county, both the eastern and middle ranges; it flows through five-sixths of the width of Massachusetts; and it prolongs its run through that portion of Connecticut which is in fact a continuation of the Berkshire region to the south. For natural features refuse to be held by political boundaries, and state-lines are not often run by geologists or their kin. And from the time it gets its license in the shire-town of the county until it passes the rugged hills below the mouth of the Shepaug, the Housatonic is to all intents and purposes a Berkshire river.

In all its characteristics it is essentially like the region through which it runs. It is not a sensational river, and its easy course is marked by no extraordinary natural features, as though nature were straining for effect. For just that reason our river is most charming. Your true lover of nature is not very tolerant of excesses and of freaks. He likes best the dear familiar things which he sees everywhere and which come to have the attractiveness of the old habits and ways of a friend. Bradford Torrey, visiting the Natural Bridge in Virginia, did not find it charming or that it "wore" well. And he confesses frankly to a preference for the common sights and scenes. "A wooded mountainside, a green valley, running water, a lake with islands, best of all perhaps (for me, that is, and taking the years together), a New England hillside pasture with boulders and red cedars, berry bushes and fern patches, the whole bounded by stone walls and bordered by grey birches and pitch-pines, —for sights to live with let me have these and things like them in preference to nature's more freakish work." This is the type of scenery through which the Housatonic runs.

The river brings to its final channel the waters of all three of the ranges of the north county—the Taconic, the Greylock, and the eastern hills; and as it flows, it gathers to itself the clear streams from scores of woody ravines and lower-lying meadows. Sackett and Ashley and Roaring Brooks out of the fastnesses of Washington, and Warner, and October Mountains; Hop Brook from the hills of Otis and the picturesque valley of Tyringham; these from the east, with the Agawam and Konkapot in Stockbridge, before the big ridges east of Barrington and Sheffield drive the streams of Monterey and New Marlboro around by way of Ashley Falls and Canaan, to find and join our river there. From the west come Yokun's River, and Williams, and Green, with Hubbard's Brook from the heights of Mount Washington township by a northerly route, and Schenob Brook, out of the same wild region by way of Sage's Ravine and the plunge at

Bear Rock. From not less than twenty-five out of the thirty-two townships of Berkshire, the Housatonic gathers its life-currents. By this sign it holds its right and title to be called the Berkshire river. And wherever in its course the lover of Berkshire comes upon it, the river seems to bear to his soul a message from the very heart of the county, from its mountain heights, its greenwood shades, its broad vales and intervales, its well-tilled fields, its vistas of enchanting scenery. Sometimes the river runs white and broken over its rocky channel like a reflection of Berkshire skies, flecked with fleecy clouds, driving before the crisp wind of the north-west. Sometimes it slips serenely along under overhanging thickets, or through grassy meadows, a reminder of the dreamy summer days when the August sun shimmers across the ripened rye. It is alive with the life of the hills.

Modern civilisation, which is hostile to all grand natural features, to forests and to mountains, to waterfalls and to shade-trees, seems to bear a special antipathy toward rivers. For it attacks them in every conceivable way, their resources, their utilities, their beauties. The hour in which the modern man settles beside a river is a bad one for the stream, for he begins at once to tax his powers to see how he can destroy the attractions and advantages which have drawn him to its banks. He tries to tire it out with work, to exhaust it with cruelties. He strangles it with dams, and poisons it with dyestuffs, and chokes it with sewage, and stifles it in steam-boilers. He tries to starve it to death by cutting off the forest on the mountains whence it feeds itself. He sedulously kills all the fish between its banks. And still the river forgives all and tries its best to keep up the struggle for existence and incidentally to bless the oppressor with uses and with graces. Here in our Housatonic, is a noble example of how hard a river dies. It keeps up a magnificent fight against the vandal powers of the human race, as they fetter it with dams and degrade it in sluiceways and mill-ponds. It yields the service demanded of it, albeit with many a passing fury, fretting itself into foam and broken water. And just as soon as it escapes from man's clutches it takes up its old life of beauty and of blessing. At Glendale, for example, after it has been corralled in a mill-pond and pitched over a dam, it recovers itself almost instantly, and before it is pulled into the traces again at Housatonic, it resumes its placid flow, and gathers shadows from the wooded banks, and sparkles in the sunlight as if it never had been forced to dirty work and never would be again. After passing Great Barrington, too, the stream which has been compelled to sweat and strain, and scour and scrub for the whole town, resumes its fair aspect and behaves precisely as if it never had drained a sewer or fed a boiler. In its gracious and serene flow through the broad meadows of Stockbridge there is no reminiscence whatsoever of its labours at the mill-wheels in

Lee. How can one help a species of admiration for the pluck and purpose of the resolute little river, and its unswerving effort not to be beaten, not to be other than it started to be up in the fields of Pittsfield, and the lanes of Lanesboro, and the fords of New Ashford. One rarely finds a river which so persistently keeps up its character for picturesqueness and rural beauty as the Housatonic, as long as it continues to be a Berkshire stream. Like a tidy housekeeper, as soon as it finishes with the day's tasks, the work that soils and roils, it straightway smooths itself out, and arrays itself afresh, and in clean garb and with placid demeanour, asserts its dignities and its superiority to the drudgeries of existence. It does not hold itself above the utilities; neither does it forget how to be beautiful.

The Housatonic ends its career, as so many noble rivers do, amid sordid and prosaic surroundings which plainly reveal its subjection to man and his hard mastership. Between banks of coarse sedges, and over mud-flats which reek with foul substances and fouler smells, its waters mingle with the salt tides from the Sound in a brackish blend which is teased by the keels of ignoble scows, and tugs, and dredges, and mud-craft of every sort. The railway, which has crossed and recrossed it eight times since it became a river at Pittsfield, once more mounts on trestle-stilts and goes over at Stratford for the ninth and last time. The caressing arms of the land reach out for one last embrace as its waters go sliding past

Stratford Point, and then its leaping waves catch sight of the low lighthouse by day and reflect its white beam at night as they lose themselves in the broad Sound and the Berkshire river is swallowed up in the sea.

13. THE EPIC OF THE CORNFIELD.

All through the long bright days of June
   Its leaves grew green and fair,
And waved in hot midsummer's noon
   Its soft and yellow hair.

JOHN G. WHITTIER.

THE EPIC OF THE CORNFIELD.

THE rotation of crops on the farm has this year filled the field opposite my window with a thriving growth of corn. Last year it grew rye, and the year before it lay fallow. But never was it so beautiful as these tall stalks have made it, their pale green below crowned with golden tassels nodding in the breezes like the plumes of a great army. It is a large field for a New England farm, and its eight or nine acres are filled to the very fences with a crop which I have seen grow from one foot to nine and ten in height. When I awake in the morning it is the first sight that greets my eyes, its whole expanse glittering in the sunlight like the waters of a lake. At noon its soft whispers come across into my chamber, voicing mysterious messages. In the evening dampness sweet odours exhale from it and drift into the open doors, the choicest fragrance of the farm. And all night long, when the wind is up, I hear the soft clash of stalks and blades which tells of the steady struggle it keeps up against the gale.

These voices of the cornfield have gradually blended themselves into a poem, a sort of epic of the field, a series of cantos whose linked numbers tell the long story of the soil, its marvellous connection with the life of man, its challenge to his brain and his brawn, its stern exaction of his toil, its generous recompense for all that he has done for it, and the constant relation it bears to his life, his institutions, and his progress. To read the story of the cornfield is to peruse the tale of man's progress out of savagery into civilisation. It is the Odyssey of his wanderings in search of his own kingdom of power in stable society, in expanding arts, in strengthening institutions. The tale of man's tillage of the field is one of those many stories in which is shown the vast unity of nature, the working together of all things for the good of mankind and for the glory of God.

If you will listen, therefore, with open ear to the voices of the corn, they will carry you away into a far-off past where you shall see the primitive farmer breaking up the soil with his sharpened stick or his stone hoe, which later becomes developed, through a succession of related ideas, into the spade and the plough. And when one sees in fancy that early estate of the human family, he sees the beginnings of the fertility which is the guaranty of man's continuance upon the earth.

The spade is the prophecy of all those processes by which the soil of the earth is fitted for a larger fruitfulness. The hand of man upon the soil means more fruit than it could bear without his help. He makes a thousand grass-blades grow where there was one before. He multiplies every grain of corn and

The Harvest-Time, Egremont.
(Looking toward The Dome.)
" *The tale of man's progress out of savagery into civilisation.* "

wheat and rice and sugar by a hundred thousand. Man must help the earth before the earth will do all she can to help him. But when man and this earth form a partnership there is no limit to the richness of the product. There need be no fear that man will outgrow the food-producing capacity of the world, and so starve to death at last. A witless creature might overpopulate the earth; but man's brain will always outrun his stomach, and his wit will provide for his appetite.

But this thought only adds to our sense of man's absolute dependence upon the soil. He is as much rooted to the earth in his physical life as a plant or a tree. For he must have his food; and his food grows on the soil and out of the soil. The human race walks erect indeed, but it still goes on its stomach, nevertheless. Cut off man's dinner and you stop his work, his wages, his health, and his pleasures. For only think, that with all his growth and his advancement, his gain in endurance, skill, wisdom, and self-discipline, it is still needful for him, three times or so in every day, to quit his work, drop all his tasks, turn from his sorrows and his joys, to feed his body and satisfy his hunger. A man may isolate himself in the highest pursuits and sometimes forget that he walks the earth at all. But by and by his hunger will remind him that he is of the earth, earthy. Hunger is a monarch that rules the world.

In every year that passes over the great nations of the world there is always one period when, but for the fertility of the soil and its return to our hands, we should be within a month of absolute starvation. For when we approach the harvest, the world's barrel of flour is almost gone, and there are no storehouses whence it may be replenished, save the deep, rich granaries of the earth,—the cornfields of this world. Let universal drought burn these harvests that are ripening ; let floods drown them ; let mildew blight, or insects devour,—and before we could plant and ripen another there would not be left enough people to reap it and gather it into barns.

It is only as we realise the primary importance of the soil and its products that we can perceive how every other industry in life depends upon the farmer. Let him drop his spade, and the carpenter must let fall his saw, the smith his hammer, the soldier his sword, the weaver his shuttle, the writer his pen, the seamstress her needle. Every other workman on earth looks to the farmer for the work to go on or stop. If the farmer will not give him his meals, he cannot work, either for self or for others. Nay, more, he shall lack not merely the strength for his work, but the material for it as well. Think how many of the industries of this life have to do with the raw material which is the product of the farmer's toil. The flour-mill, the cotton-mill, the woollen-mill, the sugar-refinery, the brewery, alas ! and the distillery too, the baker, the grocer, the confectioner,— why these are but a hint of the vast army of people whose "job" would stop if the farmer failed to send in the raw material. Forty per cent. of the workers of this country follow the various pursuits of agriculture. In the South, they farm on sugar and cotton and oranges and garden produce. In the North it is hay and potatoes and corn and beans. In the great Northwest it is wheat and corn and oats and barley. But outside those actually engaged in agriculture it is safe to say that forty per cent. more work on the products of agriculture. The farmer furnishes at once the world's bread and butter, and the labour by which that bread and butter is earned.

When we see the looming walls of the sugar refinery in the heart of a great city and realise the tremendous power it stands for, and the influence its industry exerts upon every corner of the land, we are apt to think of it as something that rests upon its own foundations and runs by its own momentum. But there is a mightier power in the land than the makers of sugar. The farmers who grow the sugar-cane are in reality the court of last appeal. Behind the men of the trust are the men of the spade; and it is they who really hold the world's destinies. It is the soil that under-runs all industries, and makes possible the great enterprises of the world of manufacture and of commerce.

But in order to till the land, man must halt. He cannot be a rover. He must give up his nomadic habits. The herdsman and the hunter may travel to find game and pasturage. But he who makes cornfields must be a permanent resident. So when man first stuck a spade into the ground to till it and dress it, that tool in the dirt was the notice to all creation that he had "come to stay." He is a "settler" now. He belongs to a spot; and he will have the spot belong to him. The garden and the farm imply houses, homes, settled abodes. The spade that turns the sod for a tilled field digs at the same time the cellar for a permanent dwelling; and it is to the discovery and the use of the spade that you and I owe our pleasant homes.

But when we have said this much we have not begun to tell the story of the wonderful changes and advances which grew as fruit out of seed from this change from the nomad's life to that of the settler, the transition from grazing-fields to cornfields. The permanent abode meant a better house; and here was the origin of every modern convenience and luxury. For the nineteenth-century dwelling is a slow evolution, an accretion of features, all of which would have been impossible in the tent of a roving shepherd or hunter. With the farm life, too, woman begins to emerge from the position of a chattel and becomes a partner. She begins to divide the labour more evenly with her mate, and to share more equitably in its gains. This, too, is the introduction to the land question. The whole problem of the soil, its ownership, its taxation, rent, tithes, titles, and deeds, begins with the settlement of the early man into a farmer. So that the real-estate broker really owes a great deal to that first farmer, Adam, who, though he is thought to have done much mischief by his experiments in horticulture, nevertheless has conferred great good by his husbandry.

The turn men took to farming had another and remarkable result. It was the beginning as well of organised and industrial society. The shepherds and the hunters are wanderers, nomads,—more than that, —they are fighters. They must drive away those who possess the land in order to take it for the game or the pasturage it affords, which they desire. The farmers, once in possession, are peaceable. They want permanence and quiet. All their influence is for order, security, peace. From the time man first began to grow corn, the relative proportion of human strength applied to war has been less and less, the part given to productive labour has been more and more. So that all the power of the farmer has been to make men more and more sociable, less and less suspicious and hostile.

But that was not all. The farmer gave the first great impulse to a division of labour. He must sow in seed-time, and reap in harvest, make hay while the sun shone, and take advantage of fair weather. So he must have others to build his houses and make his tools, and spin and weave and bake for him. The spade helped wonderfully to divide men into groups of diversified labour. It was the starting tool for the great, complicated industries of modern life.

These meant, of course, a great impetus to commerce. When many men had many things to exchange, and when the farmers had food to give in barter, the way was open for a peaceful way of getting a living. The man of the spade was the man who helped trade. He did still more. He was the means of introducing a new political principle. Hitherto the only conceivable bond of political combination was that of kinship. Men organised politically in the family or clan. But now they had interests in common because their lands lay near one another and because they had common occupations.

But the farmer-type, the industrial kind of man, was bound to be the ablest, the strongest, the fittest to survive. Tribes which had learned to till the ground and rear flocks were the best fitted to rule their neighbours, for the simple and sufficient reason that agriculture lets a vastly greater population live in a given area, by giving them plenty of food, won by tillage. The agricultural tribes were drawn into communities, villages, towns, cities. They grew socially. They associated, co-operated, waxed strong in resources. They prevailed over their enemies. They not only proved that in union there is strength; they showed the natural and inevitable advantage of having plenty to eat and wholesome homes. They were living witnesses to the grand truth that the very soil of the earth itself is in league with progress, and peace, and higher ideals of life. They stood for the rise of man, the coming of this child of God into his higher inheritance, the wanderings of this wayfarer among strange scenes toward the real home of his soul, in an orderly society existing for the highest ends.

That is the epic of the cornfield! That is the wonderful tale, so full of the deepest and most passionate life of man upon the earth. What wonder that these multitudinous stalks move in their places like the surging of the human beings who have shared in this great upward movement of progress to civilisation! What wonder that the voices which they utter fall upon the ear like the marching-song of a world advancing to its proud triumphs in the industries of life, its arts, its statecraft, its religion! The waving cornfield is vital and vocal with the grand life-story of man.

14. THE SEAMY SIDE OF SUMMER.

    And all the land
Lay as in fever, faint and parched with drought ;
And so had lain, while many a dreary day
Dragged the long horror of its minutes out.
         JOHN W. CHADWICK.

THE SEAMY SIDE OF SUMMER.

THE sun is just going down in a splendour more wonderful than the glory of kings. Over the close-cropped ryefield and the maples beyond, and the hills behind their green domes, shines a purple light in which every tree and every stretch of meadow seems transformed into some dreamland scene, and throbs with an unearthly beauty. The sun's disk is deepened to a dull red from which all the gold has been extracted, and when it is fairly below the verge the light and colour fade at once and leave the landscape to the fast-gathering grey of the dusk. It is a scene which makes the observer admire reluctantly ; for he knows that every hue and every glowing effect is born of the earth's pain and unrest. This purple flush is the hectic of the drought-fever. It would not redden the west but for the heats which now for weeks have parched the fields, and dried the soil and the air, and burnt the life out of the herbage. It is unnatural and unwholesome, and it is a reminder of the seamy side of the summer.

I have sometimes thought that there is a visible reluctance on the part of nature-lovers to describe or to refer to this trying side of the outdoor world in summer. The disposition is not strange. We do not like to dwell upon either the faults or the misfortunes of our friends. He who loves nature likes to think of her at her best, in her phases of beauty and grace, when she exhilarates or inspires. To tell of her pangs, to paint her in her distresses, to show her in poverty and loss and disaster, seems to savour of disloyalty or indelicacy. Perhaps that is why we hesitate to tell the story of the drought.

Yet never does human sympathy seem to come so close to nature, never does the heart of man so yearn toward the dumb earth and all that it bears, as when the sun is parching and powdering and cracking the soil, when the heavens refuse their moisture, and when the very air is suffocating with the dust of the field. To-day has been a typical one of all seasons of drought. The sun rose in a dry fog, and all through the forenoon the hills swam in a blue haze which took all the contours and the modelling out of them, and left them mere flat shades of indigo against the sky. As the sun climbed, a few clouds tried to form; but they drifted like so many handfuls of ashes into the domed oven of the firmament. Far above them, in the remote upper regions of the atmosphere, a few faint wisps of cirrus-cloud lay motionless, all tinged with a reddish glow which seemed like a faint reflection of the seared fields and dusty roads beneath them. Thus these clouds, which usually carry suggestions of shadow, coolness, refreshment, all day have mocked the hot world and mirrored the picture of its discomfort.

That dreary picture lies all about us. The grass tells the most pathetic story. Every suggestion of the emerald spring is gone. The hillsides show as barren and sear as they ever will in November, and the long grasses by the roadside stand like so much uncut hay. The mowing-field gives no sign of a second crop, but sends up its shimmer of heat like the stubby ryefield next to it. Yonder the corn struggles to grow and ripen, but its green is pale, and its waving expanse is streaked with brown and yellow, and its blades are curling at the sides and rolling into sharp bayonet points. There is rusty brown on all the foliage, and here and there great patches of dull colour in the woods show how the merciless heat has hurried the leaves to their old age, and dragged October into July.

But it is when one goes out to meet the landscape, and comes to close quarters with these familiar scenes, that he enters into all the distress and hardship wrought by this drain upon the soil, this undue stretch of every resource in the life of plant and tree through dearth of moisture in air and earth. The July wild flowers which often linger in a thrifty maturity well along into August, have withered and dried and now rattle their seeds in pods which have garnered all that is left of their brief lives. Blue vervain and St. John's-wort and the cheery toadflax are but so many brown stalks upholding darker patches against the pale dry hay-colour of the grass and stubble. The goldenrod is a full fortnight early, and as it hurries into bloom brings the same impression of precocity which one gets from those poor children of want in city depths who have grown old in hard experience before they have done with their youth. There is something unwelcome, too, about this forwardness of the golden blooms, for they hasten the omens of autumnal change and keep the eye full of reminders of the rate at which "the roaring looms of time" are weaving their endless web. One cannot bear to have the hands on the clock of nature turned ahead. At their normal rate they outstrip all our estimates. When they thus discount changes and times and seasons to the eye, they seem to cheat us of some actual fee-simple in the days.

The way to the woods is but a short one, yet now that the dust is thick in the road, and lies like a dry frost over the burnt turf along the fences, and makes conspicuous the cobwebs bespreading the grass-tops, it is an uninviting path. The eye recoils from the dreary monotony of dust, and the feet slip painfully over the worn and polished carpet. One is glad to reach the wood-road and turn aside from the open, and welcomes the promise which these shady coverts hold forth of a respite from this sense of exhaustion and of thirst. But the woods have their story to tell. The trees and the sturdier shrubs are weathering the trouble well; but mark how the ferns are yielding to the strain. The *Onocleas* are yellow and shrivelled, the *Osmundas* are turning brown and crisp and curling into a shapeless shrivel. Even the sturdy brakes, the *Pteris Aquilina*, are giving up the fight and dying by dozens. The great bog running through the woodland is as dry as a brick-kiln. The cardinal-flower blooms there as a matter of habit, and the purple-fringed orchis. They show the ravages of the dry term less than almost any other flowers. But all the soil where they flourish has lost its rich and steamy smell, dank and heavy with the ferment of woody soil and moss and fern. Only the mosquitoes thrive undaunted, and hum with as sharp and strident a twang as ever and bite as merrily.

The eye falls on the brilliant fruit of the trilliums, and the clumsy leaves of the hellebore, and the vision of the spring comes up before him, when all this soil was soaked with the melting snows and drenched with the copious showers, and the earth was bright in emerald greens, and juicy with the succulence of tender verdure. And now the earth cries in vain for one poor draught from a passing shower, and the clouds strive fruitlessly to rally for a storm, and the east wind is as hot as the wind of the south.

And how helplessly both nature and man lift up their varying appeals for respite and for aid! How oblivious seem the powers of the air to all this necessity! How impatiently we await the rounding of the mysterious cycle which must complete itself before the skies shall yield their reviving store, and the great storms travel from the west and from the south, and the soil drink deep and slake its awful thirst, while every parched and needy herb laughs in a long-deferred glee.

15. FRUITFUL TREES.

I shall speak of trees as we see them, love them, adore them in the fields, where they are alive, holding their green sunshades over our heads, talking to us with their hundred thousand tongues, looking down on us with the sweet meekness which belongs to huge but limited organisms.

O. W. HOLMES.

FRUITFUL TREES.

I HAVE been out under a tree all day. The winds have talked into its leaves, and the sunlight has flashed its message through the quivering foliage, and I have listened with straining ears to catch, if I might, the wisdom which the venerable sages of the wood were gathering from the breezes and the sunshine. I am sure I must have interpreted somewhat of all that was said, for there came to my mind a long story, drawing my thought out toward the trees, and bringing home to my consciousness things I had never thought before. I was borne forth in fancy to the homes and the cities of men, to their busy, reckless, tireless, intense, and pushing life. I could see, by a flash of intuition, how closely the life of the trees, their functions and their influence, are linked and bound to the life of the whole world and of man himself. I realised as in a vision how near the trees lie to the fountain-heads of those streams of power and resource which bear to man his largest riches.

I do not mean just the obvious uses of the trees after they are slain and harvested into men's lumber-yards and sawmills and planing-factories. Everybody can see how they have contributed to the building of our cities and the construction of our homes. We in America attribute the rapid growth of our population to our unoccupied lands, to our ocean ferries, to our spreading railways. But we may not forget that the railways of this country could not have been laid with the rapidity which has marked their construction if the trees of our forests had not furnished the ties on which the miles of steel rails have been laid. We talk about our great protected iron industries as the source of our rapidly developing railway system as if we had somehow made it with our legislative devices, our far-sighted statesmanship and laws. And we never bestow a thought on the munificent contributions to the great result made without a device or an effort upon our part through the bounty of Him who clothed our lands with forests. Take out the wooden ties which hold the rails in place, and you strike out at least one-half the facilities which have aided us in so quickly opening our vast territory to the use of the world.

Even now, when ruthless consumption of our forest resources has compelled us to think a little about the future, we are sobered and anxious as to how we are to maintain the lines we have built, much more to build new ones, if the ravaging of our forest-lands goes on. Moreover, how could we ever have housed the millions who have come to us, if we had depended upon the slow processes of quarrying stone and baking brick, and had not been able to give temporary shelter to the incoming throngs with frame houses first. The development of the great West would never have been possible but for the mighty forests of Maine, of Michigan, of Wisconsin, of Minnesota, of the States on the Pacific Slope. The waving branches of the trees of our woods have beckoned a welcome to every immigrant upon our shores.

But there is a closer relation yet between our prosperity and our trees. The forests as they stand, the trees growing and working, the trees as they exercise their natural functions, are more valuable friends than in their post-mortem estate. They have a most vital relation to the fertility of our soil, to the state of our climate, to the perpetuation of our streams. For nature has made the tree one of the great conservators of the soil. Out of the deeper layers of surface soil, out of the circulating air, the trees imbibe rich chemic gases, which they build into solids and deposit in leaf, twig, and trunk. So that when these fall and moulder they lay upon the top of the ground new harvests of nutrition which they have reaped in the fields above and the fields below. So that all the trees in the valleys are heaping up enrichment of the soil where they stand ; and all the trees on the hillsides are doing the same thing and more.

For every rill which flows down a hillside bears some of the earthy salts which feed the soil with new life into the vales below. When our steep mountainsides and hillsides are covered with forests the valleys are sure of enrichment. The soil will not grow poor. The trees on the heights will furnish dressing and furnish it for nothing. But cut away the trees and the soil has lost its feeders. It grows poor, thin, and meagre. The fields below are equally losers. They have been deprived of their best friends. There is many a region in this country where the hills and the mountains have been denuded of their trees, which is now too barren to do more than support a meagre crop of huckleberry bushes. The secret of this poverty lies in the fact that the settlers and their farmer-descendants have squandered their woodlands ; and he who wastes his fruitful trees lays waste the very soil beneath his feet.

He commits a blunder even worse than this. He is accessory to the destruction of that soil altogether. He opens the door for the entrance of two foes of the harvest,—drought and freshet. The trees are our great defence against these two enemies of fertility and abundance ; and singularly enough the two are exactly opposite in their nature,—excess of wet and excess of dryness. Woods and their undergrowth are man's only protection against inundations, and the only means by which these floods held back can be stored up for distribution through whole seasons. The reason, however, is simple enough. Let the hillsides be wooded, and the roots of the trees, with the soil they accumulate, the mosses, and the vegetable mould, make a spongy, oozy mass, which holds the falling showers and the heavy rains, and lets the gathered floods trickle slowly through, down to the thirsty lowlands. The surface underneath may be nothing but bare granite ledges. But as long as the trees lock roots, as long as they cling to the crevices and the ridges, the soil will stay there, anchored by the guardian trees. Thus the moisture which the clouds distribute will be husbanded by these thrifty friends of our agriculture, and carefully dealt out to the rills and the brooks and the soil below in the valleys, little by little, all through the hot summer.

But cut down the trees, clear the hillsides, and see what happens. The thin soil, no longer protected by the trees, no longer held in place by their netted roots, no longer shaded by their leafy branches, grows dry, and crumbles, and loosens. The heavy rains wash it bodily into the valleys. The bare ledges appear. The vegetation dwindles. The hill or mountain becomes a barren crag. Its brooks and springs dry up as soon as they are filled. The drench of the hillsides is hurried in bulk down into the valleys; and every rain-storm becomes a swift freshet, destroying the crops and threatening house, barn, and factory, at the same time that it washes down the sand and gravel from the heights to deaden and impoverish the lowland meadows. But as soon as the rain stops, the streams stop too. They dry up and shrink in their beds. They disappear under the scorch of the sun. The same fields which were inundated in the springtime are parched and dusty in the heats of midsummer. That is the way we are enriching ourselves. We are paying dividends at the sawmill, and putting mortgages on the farms. We are burying our fields at the same time that we are destroying our forests. Nay, more, we are sapping the sources of the very water-power which runs the sawmill which cuts up the trees.

There are factory towns in New Hampshire and Massachusetts whose prosperity is seriously affected by the lowness of the rivers and streams in midsummer. The reason is to be found back upon the mountainsides, where the fast-thinning ranks of the forest-trees show why it is that the floods in April and the droughts in August make such havoc with the profits in the factory counting-rooms. It is becoming a very vital question to the American people whether they will suffer themselves to be exposed alike to drought and to flood through the reckless robbery of the mountains and the pillage of the great forests of the land. The time is coming when the safety and the preservation of the lordly river which debouches past New York to the sea will have to be decided by the most vigilant care of the Adirondack forests and the most unremitting warfare upon their foes. Ignorance and greed always stand, axe in hand, ready to transform the trees into logs, and coin a dollar to-day, though it cost ten for damages to-morrow.

This fact about the trees teaches us a lesson which is always timely. These bare crags where once were

Fruitful Trees.

waving tree-tops, looking drearily down on the dry rocks once laved by abundant waters, proclaim the short-sightedness of purblind men. Homestead and farm in the valley and upon the plain are bulwarked and defended by the far-away forests which lift their faint blue above the horizon. Yet the farmer and the cottager give no heed to the spoiler of the forests. So long as there is a fire on his hearth and a harvest in his field, the dweller by the river's bank will not sit up at night to watch lest the hosts that protect his fields are slaughtered at their posts. That is so thoroughly human! We are so slow to trace the connection of what is a little remote from our daily sight and hearing, from our present consciousness, with our personal comfort, safety, salvation.

It is a hard thing to make a man believe in the importance of anything which does not press immediately and heavily upon him, threatening him with loss or proffering him gain, inflicting pain or affording comfort. It took ages to teach man that it was profitable for him to provide for a meal or two ahead of his actual necessities. It took the practical demonstration of steel and gunpowder to make the American people believe that slavery would cost more than it yielded. And to this day every child in our homes has to be taught afresh the lesson that it is not prudent for a moment's present pleasure to invite a week of future pain.

Or, to go a little deeper, these friendly trees teach us that we ought to learn and remember the remote and retiring sources of our actual life. A man is fed and sustained by a thousand helpers of which he never stops to think, that get none of his thanks. The great streams which float our commerce and turn our mill-wheels and irrigate our lands have their source and supply high up in the still mountain forests, where there is no hint of traffic or industry or growing crops. The stream of dollars which runs southward through all the industries which line the Mohawk and the Hudson, has its rise in the Adirondack forests, and the green woods of the Catskills; and when the tellers of the New York banks catch it in the reservoirs of their bank-vaults, they are really storing up the product of the trees of the New York forests. Yet how little the merchant and the banker, and even the manufacturer in Little Falls and the farmer in Dutchess County, think of their dependence on the mountain woodlands, or how hard it would go with them if these were swept away by fires or by the axe.

Precisely so we ignore the real sources of our riches, our prosperity, our safety, our faith. Every farmer must have his almanac; but when he goes to the legislature how hard it is to get an appropriation from him for the astronomer and his observatory where almanacs are made. Untold fortunes have been coined out of the telephone and the electric light; but how much of that wealth has ever gone back to the schools and the laboratories and the college classrooms where, by slow and patient study and experiment, the mysterious fluid was broken to harness and set to work? Men question the value of studies which are disclosing the very paths of prosperity and of social health. They deem the germ-thoughts, the theoretic seeds out of which the most conspicuous advantages ripen for man, impractical and worthless. They are impatient of the doctrines which lie at the very fountain-head of the world's best life, and have no sympathy with men's efforts to maintain or to find out truth, so long as it does not directly appear how the truth can coin a dollar or relieve a pain or fill an empty stomach.

But I am wandering from what the winds and the sunbeams told the trees, and making this humble stump which has served me as a chair-back a veritable pulpit. Let these words be numbered now, lest they quickly drift into doctrine and bring reproach upon the trees. Heaven knows they suffer enough at the hands of men; would that this discourse might help to create a little more care for them in a wise way!

16. THE WINGS OF THE WIND.

The mountain-wind!—most spiritual thing of all
The wide earth knows.

W. C. Bryant.

THE WINGS OF THE WIND.

ALL day long the voice of the gale has filled the air with its hoarse call. Last night the sultry clouds of the thunder-storm hung low in the stifling air, and it seemed as if the earth were too exhausted ever to do more than gasp for its poor breath. But when the dawn began to glimmer in the east there came a puff from the northern hills, the leaves began to rustle in a lively dance, the gust became a steady breeze, the breeze grew to a gale, and the gale sometimes got excited and lashed itself into squalls which set all the trees of the wood to tossing and writhing in wild struggles to hold their own against this reckless, riotous, roystering blast. The oatfield at the side of the house has been one weltering sea of green all day. The corn in the field beyond has been a tatter of green streamers fluttering down to leeward. Every wooded hillside has been a billowing mass of shifting greens. And all the while that mighty voice has roared in our ears until they are fairly tired with the strain of enforced listening.

So all these bright hours, bracing the body with the tonic airs of the north, have filled the mind with thoughts of Him "who walketh upon the wings of the wind," and of this wonderful messenger He sends upon His errands of mercy. Of all the phenomena of nature, that which is sealed the closest to most men and women is the story of the winds. We realise less of the wind's office and function than we do of the work of any other natural agent. It is as true of our day as of the day of Nicodemus, that "The wind bloweth where it listeth, and ye know not whence it cometh nor whither it goeth." This great envelope of air is all the time pressing on us, moving by us, bearing to us the very breath of life, and yet it is probably the least understood, the least appreciated, of any of the great natural forces or facts. The wind which blows across this earth is only the air moving about on its errands and performing its work. But when we search out what that work is, we realise how true is the word, "He maketh the winds His messengers."

Think what this atmosphere of ours is, whose touch we hardly recognise save when it brushes against us in the hurry of its toil. It is the breath of life to man and beast and plant ; and by a curious law of economy the waste gases which the man exhales are the food of the plant, and *vice versa*. It is one of the minor services of the winds that they are continually mixing these in their due proportions and maintaining the salubrious equilibrium of the atmosphere. The winds moreover bear the moisture which the air acquires in its wanderings over the sea to the mountain

The Edge of the Storm, Egremont.

(Looking across Fields toward Hillsdale.)

"It came rolling down upon our plain to belch wind, rain, and hail upon us."

heights, and sink it into the springs which nestle there. And then, as these dried-out currents pass down over the arid plains, the "wings of the wind" bear them to moister regions to saturate them afresh with the quickening drops of water. The winds set in motion the vast currents which are flowing in the deeps of the sea, and so add another to the great climatic agencies. They do as much for the shallower waters. It is the winds which are carving away the shores of Cape Cod and Coney Island by setting on the waves to do their work of destruction and of change.

The winds may well be likened to the great transportation systems which man has created. What railways, rivers, canals, and the ocean are to human industries and interests, by means of their great trains of cars, their boats, and their steamships, the winds are to the natural world. They are the grand vehicles of exchange. They bear the products of every zone to every other zone. They warm the poles with the airs from the tropics. They cool the tropics with airs from the poles. They transfer the seeds of a thousand grasses and plants to new fields, and sow the desolations with verdure. So also they carry off the superheated airs of the earth's surface on the "up-tracks" and bring them back cooled, cleansed, revitalised, in the winds which blow down from the higher altitudes after the thunder-shower or the storm.

It seems to me that in this law of the winds we are permitted to read a startling lesson of the Divine

Plan to unify all lands. The winds, far more than any work of man, bring the ends of the earth together. They are the great levellers of barriers. They are the democrats of nature. There are no walls they will not overleap. They recognise no distinctions of race or of rank. They carry their boons with unswerving impartiality. They bear their desolations with unsparing vigour. They turn this earth into a very small place. They make neighbours of the most separated coasts.

Before the storm has crossed the Mississippi River, the winds have borne the white plumes of the cirrus clouds, the forerunners of the gale, to the lightship off Sandy Hook ; and the warning signal flutters for the sailors going out to sea. The winds pick up the dust of a volcano in one continent and drop it on another. They bear the airs of the equator to the arctic circle. They waft men's ships from end to end of this globe. They carry the germs of the grippe "around the world in eighty days." But just as willingly they bear great clouds of pollen or the seeds of innumerable plants, to scatter them in new lands, and cause new crops to grow in waste places. They know no east, no west, no north, no south. To them the world is one neighbourhood. The winds are a great natural illustration of a law of God. All nations are as one village. All mankind is one brotherhood. While men are building their Chinese walls of one sort and another,—creeds, political platforms, state boundaries, tariffs, tongues and languages, forts and navies,—God Almighty is sending out His winds as messengers to proclaim the kinship of races, the neighbourhood of nations. We cannot resist His will. We are destined to see our boundaries wiped out, our narrowness neutralised, our provincialisms annihilated, by Him who hath made of one blood all nations of men.

For He has provided for a circulation of ideas which is just as real, just as free, just as effectual as the circulation of the winds. He takes care that the truth uttered in America shall cross to Russia. The principle of government embodied in an institution in the heart of Asia reappears in the depths of the German forests, in the Parliament Houses at Westminster, in the New England town-meeting. The word John Wyclif uttered in England, Luther hears and passes on to the world. The gospel proclaimed in Judea is echoed to the isles of the sea. The winds of truth blow where they list, and no man knows whence they come and whither they go.

Just as He makes the winds His messengers, He will take anything else that comes to hand to serve His end. It was a little ship which crossed the seas in 1620 and brought to Massachusetts Bay the seed of a new society, a seed ripened on the old soil of Holland and of England. It was a little book by Harriet Beecher Stowe which was the message of enfranchisement to American thought on the subject of slavery and which led to the enfranchisement of the slave himself. The great liberator himself was lowly of birth and came out of humble station to his great task of national service. Nor does God insist on voluntary service ; He makes even rebels and sinners do His bidding and become unconscious and reluctant co-workers with Him. Pharaoh was His unwilling coadjutor. So was George III. So was Jefferson Davis. Doubtless Tammany Hall may yet turn out to be the greatest reform movement this country ever saw, and work out by the rule of contraries that purgation of municipal politics which it is its own chief aim to hinder and to prevent. It must be a very skilful agent which can outwit omniscience and escape carrying God's messages and bringing His will to pass. Says the Concord poet-sage, making God the speaker:

> "My will fulfilled shall be,
> And in daylight or in dark
> My thunderbolt hath eyes to see
> Its way home to the mark."

But if the winds illustrate the process of human unification, so they do also of another of the great laws by which the labours of God go on. I know no finer instance than they afford us of that great law of rhythm, which characterises the whole creation, which contains some of our most encouraging hopes, which throws a light on the methods of the Creator, and the plan of creation's unfolding life.

You and I, my reader, have our most frequent interest in the winds in connection with the storms which come down upon our coast. And there is a peculiarity of these storms which everybody is sure to notice at some time or other. Who has not observed how the stormy day falls week after week upon a Friday, or a Sunday, or a Monday? The weather often rolls along in a succession of rhythmic movements as regular as the surges of the sea. So we have come to talk familiarly of "warm waves" and "cold waves." The rise and fall of the mercury in barometer and thermometer are visible signs of what goes on in the atmosphere. Our storms are literally waves of the air,—disturbances which throb with a rhythm like that of the waves of the sea. The laws of these winds are as fixed as those which govern the ebb and flow of the tides, the revolutions of the seasons. First a wave of clear skies and dry winds; then one of damp breezes and rain and clouds. There is a throb to the atmosphere like that of the sea, the seasons, your heart or mine.

It is a phenomenon which is part of a great universal law,—that all motion, all life, moves in rhythm, in undulations, in throbs, waves, periods. It is shown in every one of our bodies. Beat of heart, inhalation of lungs, stomach-hungers, sleep and work, pains and pleasures, all show the inevitable law of rhythm. Life itself is a succession of periods of joy and sorrow. It moves in waves. Our moral life fluctuates. Bright days follow depressed ones; virtue is now hard, now easy; hope and despair alternate; no man has a "level best." There are revivals of religion. When men sneer because the moral crusade of Francis Murphy or of Dr. Parkhurst is "only a wave," they forget that "waves" are all that clear our atmospheres, and that all great changes in the world move in just this way. What else but successive, wave-like epochs of heat and cold, upheaval and subsidence, have made the world what it is to-day ? How has intellectual life grown but by periodic intensifying of its activities,—now in the decadence of the Egyptians ; after a long depression, rising in the philosophy of the Greeks ; in the revival of learning in Europe ; in the scientific wakening of the eighteenth century ?

How else has civilisation advanced ? We see the Indians disappearing before the march of an enlightened race. It is an old story. Wave after wave of tribes and races has rolled westward from Asia's heart to the Golden Gate ; kingdom after kingdom,— Assyria, Nineveh, Babylon, Egypt, Greece, Rome, the mighty children of the North, the Teutons and the English. Who knows what mightier, better, holier race shall whelm our own, and teach a nobler civilisation ? Even the kingdom of the Christ is to give place to that other,—when " God shall be all and in all."

So when the wave of storm and rain blows down upon you in the wings of the east wind, or of the wind of the south-west, and then the "cold wave," the great undulation of clear weather follows after, bare your brow reverently. You are witness to a mighty pulse-beat of that endless rhythm which began when God said, "Let there be light," and the evening and the morning,—the alternating throb of light and dark, the diurnal wave of earth's life,—"were the first day." For the "stormy wind, fulfilling His word," bears the message to all the corners of the world. "Let every thing that hath breath praise the Lord."

17. AT THE SIGN OF THE BEAUTIFUL STAR.

Lone watcher on the mountain-height,
   It is right precious to behold
The first long surf of climbing light
   Flood all the thirsty east with gold.

J. R. Lowell.

AT THE SIGN OF THE BEAUTIFUL STAR.

THE French have a charming phrase to describe a bivouac in the open air. They call it "coucher à la belle étoile"; and the expression treats the starry firmament as the inn where the traveller finds shelter for the night. It is a poetic and suggestive reminder of that ampler roof which covers us all,—the great mansion of the universe which houses all our lesser worlds and homes. Sometimes in a man's life it is worth his while to act upon the hint of this happy phrase and put up for the night at the sign of the beautiful star. Every year the passion comes upon me afresh to go out and sleep under the open sky, to feel "the sweet influences of the Pleiades"; to be lulled by the night-wind; and to renew in my soul that sense, which comes only to him who from earth's high places sees the moon swim past him like a sister ship in the fleets of the firmament, that I myself am a mariner in space. Thus it befel that on a glorious afternoon in August I succeeded in beguiling four companions to accompany me up the Dome of the Taconics, to sleep,—or wake,—under "the beautiful star."

It cannot be too often insisted that he who would

enjoy a mountain must pick his day; and he who would enjoy a mountain at night must be doubly careful in the selection of the night for his pleasure. It is better to wait a month for just the right turn in the weather than to select a poor time and waste a night in the fog or in the rain. The fit time is such a one as we had chosen. Heavy thunder-storms in the night had left the air quite clear of vapours, and the north wind, still blowing freshly at noon, was the assurance of a clear evening and morn. At three o'clock it was decided that the time was ripe for the long-projected bivouac, and at half-past five we were at the mountain's foot with our packs of food and blankets, moving up the easterly slopes of the forest-clad pile.

It is a peculiarly choice experience to climb the eastern side of a wooded mountain in the late afternoon. One enjoys then a prolonged sunset effect. The sun is already below the mountain's crest, and the shadows are gathering in the forest which he threads. The silence of the mountain solitude is broken only by the vesper song of the thrush. And the stillness and cool shade fill the climber with premonitions of the intenser silence and darkness which await him on the summit to which he moves.

The sun was a half-hour too soon for us, weighted as we were with our loads, and by the time we had reached the mountaintop it had been fully fifteen minutes below the horizon. The twilight was fast fading as we prepared our simple camp upon a flat ledge, some five minutes' walk below the summit, with a fine eastern prospect. A little "wind-break," made by stretching a rubber blanket between two trees and slanting it back to the ground, was all the shelter we had; and the surface of the ledge, upholstered scantily with grass and moss and wild plants, was our very primitive couch. But when the campfire had been kindled, and had begun to throw its flickering light into the shadowy wood, and to flare and flash out into the great empty spaces of the night, it seemed as if we had created here in the forest, and on the solitary height, a charmed circle of cheerfulness and security. And over all the stars glittered in a stupendous mockery of our tiny torch, but with an air, nevertheless, of infinite and tender protection.

Such starlight, indeed, it has rarely been my lot to see! The sky was as clear as on those cold nights in midwinter of which friend Hosea Biglow used to say that they were

"All silence and all glisten."

A cloudless air, without a trace of haze, permitted the eye to sweep the whole firmament, and to see every star that ever made itself visible to human sight. Even after the moon had risen, the brilliancy of the stellar display was hardly abated. Through every hole and crevice in our leafy thatch there was a star peeping down upon us; and the gaps in the sparse pine and maple growth showed the great constellations wheeling in silent glory through the heavens. We followed all their splendid evolutions. When we lay down and began to stare at them, the Great Bear was climbing down the west, and when we arose to greet the dawn he was hunting his breakfast in the regions eastward.

We saw Cassiopeia's Chair turned slowly upside-down, and the Swan swim half across the sky, with Pegasus in full chase. Before the dawn began to streak the east, the Pleiades and Hyades had appeared in the south-east, and behind them came Taurus and Orion, seldom seen by summer star-gazers, because they rise so early in hot weather. It was a glorious experience to occupy this lofty observatory and fancy ourselves back in the primeval ages—mere cave-dwellers, or Aryan shepherds, looking out upon these constellations which have changed never a whit since those far-off days.

The task of the imagination was not so hard, in transferring us so far backward in time. No troglodyte ever had a ruder couch. No shepherd on the Asiatic plains ever slept more frankly under the skies. The mountain herbage was not as soft as a spring bed, and the log we used as a common bolster, even when softened by the thin folds of a coverlid, was by no means downy. We lay down in a row, each man with his hat and overcoat on, and drew a thick quilt over the entire squad. Packed thus in close order, even when the fire burned low, and only one weary eye of live coal glowed and glared at the watchers for dawn, we were warm enough. The night air had no frosty bite. It was as soft as it was clear. And when the fresh gusts rustled the leaves and shook the branches of our sylvan roof, the breath of the wind was quick with the odours of the forest, and electric with the ozone of high altitudes.

It is not to be supposed that we slept and dreamed away those precious night-watches. Does anybody imagine that we would climb four miles and drag heavy loads of food and bedding with us, and lie down on the thinly disguised side of a rocky ledge, just for the sake of getting a night's rest? We could have done this at home; and found softer beds, easier pillows, and more persuasions to slumber. We had ascended this "hill of the Lord" that we might revel in the glories of a summer night. We could sleep any time. But such nights are rare, and rarer still the chance of spending them in the open air. The price of a few hours' sleep is a small premium to pay for such a night's joy and refreshment.

The hours rolled swiftly away. There was occasional chat and banter. But for the most part long silences, snatches of sleep, open eyes gazing into the starry deeps, ears sharpened by the silences for the faintest sounds of the night. The ear had the least gratification of all the senses. In the early night-watches I heard but the voice of the wind, the "cheep" of some waking bird, the baying of a hound in the valleys, and once the distant whistle of a locomotive; later the wind died away and a profound silence reigned on the mountain,—a silence so great that I heard the leaves rustling to the ground, and the chafe of the twigs as the trees swayed in occasional drafts of the light wind.

The first faint light of the morning brought us all to our feet and out upon the ledge to greet the dawn. The broad valleys of Berkshire lay beneath us, veiled in grey lights which deepened here and there into darker shadows and the uncertainty of misty draperies. The air was absolutely still. One could hear the crowing of the cocks in the farmyards below, and up from the woods at our feet there rose the matin-song of the thrush. The grey brightened the straw-colour, and then deepened into rosy red, as into stars grew pale and faded one by one. The hills of Berkshire seemed to swim in lakes of mist, whose tides rolled up and broke against the rocky sides of these scattered ranges. As the light grew a freight train went creeping along the Housatonic, leaving a trail of smoke and steam lying almost motionless in the still air for a full mile behind it.

We sat in silent wonder at this marvellous transformation of the earth with the breaking of the day. What process or episode in nature is so astounding, so impressive, so awesome! The change is so gradual yet so swift; so gentle yet so resistless; so unobtrusive yet so majestic; so imperceptible by moments, so absolute in the space of a few short hours. He who sees the day dawn sees the sublimest spectacle in nature. We who assisted

Lenox Road and Stockbridge Bowl, Stockbridge.
(Looking toward Lenox Mountain.)
"*A road which gives as romantic views as any in Scotland.*"

at this dawn felt as if we had celebrated a grand religious rite.

With the approach of the sun, we climbed back to the crest of the mountain, and, on the open ledges of the Dome, we awaited the appearance of the crowning glory of the hour. At last it came. Out of a sky so clear that there was scarce a sign to herald its uprising, from behind the clear-cut silhouette of the eastern hills, there suddenly blazed a rim of flame, which grew into a segment of a circle, and at last into a glittering disk. Our vigil with the heavens and the silent earth was ended. When the horses of the sun come stamping and champing to the door, one must pay his charge at the sign of the beautiful star and fare forth for the journey of another day.

18. BY THE BLITHE BROOK.

I chatter over stony ways
  In little sharps and trebles,
I bubble into eddying bays,
  I babble on the pebbles.

. . . . . .

I chatter, chatter, as I flow
  To join the brimming river,
For men may come, and men may go,
  But I go on forever.

ALFRED TENNYSON.

BY THE BLITHE BROOK.

I COUNT a passion for brooks among the most deep-seated and intense of my loves. From earliest boyhood the gurgle of a running stream has always filled me with the same joy, though in a different key, as the roll of the surf or the wind in the forest trees. To come upon a brook as it slips under the highway bridge ; or as it loiters across the open meadow ; or as it tumbles over its rocky bed on the hillside ; or as it gladdens the stillness of the woods with its musical ripple,—was always a refreshment to the spirit like that which its cool waters give to the body. To no verse in the Bible does my heart yield quicker response than that which says, "He shall drink of the brook by the wayside, and shall look up." With body and with soul I say "Amen." I never pass a brook without a "Te Deum." And I have given myself a dyspepsia many a time, in a country where brooks abound, because I could not resist the seductive tipple they afford.

So it has been a keen delight in these summer days to find myself the near neighbour of a most engaging brook ; a brook which was active and cheerful all summer long ; a brook which was typical of all that a brook ought to be, outside of a forest country; a brook which from outlet to source invited human companionship and lent itself to human comfort.

We made acquaintance with it at its outlet, where it loses itself in the lake it helps to feed. It was one of our chief pleasures on hot July days to row between the beds of horse-tails and bulrushes which mark its entrance, between banks where crowding alders and willows screen the meadows on either hand, and where copses of the jewel-weed and loosestrife gleam against the green, under the branches which meet above its yellow waters, until our skiff grounded on its pebbly shoals. These brief voyages were always enlivened by sweet bird-voices, the robin, the song sparrow, the unmelodious catbird, and the clattering kingfisher all lending their notes to protest against our presence. Halfway up the navigable portion of the stream there was a break in the thicket, a pathway, and a landing-place. Through this portal to the meadow, one caught frequent glimpses of the haymakers, while the fragrance of the new-mown grass enriched the breezes; and in the near foreground rose the green urn of one of those noble elms, the glory of New England's fields, peerless in any land for beauty and for grace.

We met the brook again where it parallelled the road for which centuries ago it had surveyed and prepared a bed. The frequent rains of this moist summer had kept it full, and its voice was clear and jubilant as it hastened toward the silence of the lake. A sharp clamber brought us to a little plateau where the hilly stairway broadened into a sort of landing, and the brook stayed its swift currents in a tiny mill-pond by whose brown pool an aged mill bore witness to the ancient service of the stream; and two nestling cottages, homes of three generations of New England farmers, suggested the long story which might be told of the brook's alliance with human interests, and the parallel stream of life which for a century has run beside its own.

It was on a hot morning, when we sought a cool retreat, that we turned aside a little distance higher up the stream, and by a diverging road crossed a rude bridge; and here the little gorge through which the brook was running, with its large, grey boulders and its arching trees, beguiled us from our purpose of a longer stroll, and led us down the banks, close to its mimic flood. Here we held communion with the spirit of the brook. Here, too, we played the simple game of "tuning the brook," by damming its waters where they rush through some narrow crevice in the rocks, and drawing from the remonstrating stream a new musical note. Still strolling up-stream, over rocks and little ledges, we found within a few rods of our first halting place another shady nook which revealed a totally new aspect of our stream. That is the charm of a brook. It means variety, change, new vistas and new phases of nature along its whole course. It is the analogue of life. It is human experience foreshadowed in lower nature. And under a spreading tree we talked of the deep things of life, and opened Emerson's poems, and took fresh inspiration from his pages, fraught with a sense of the divine in nature deeper even than Wordsworth's. An hour later we again moved on, this time making a stage of at least forty feet, when we found a sunlit pool where we laved our feet, while the minnows played about them, nibbling furtively at our toes, and the big turtle, whose hole was under an adjacent boulder, came out at intervals to get his breath and to scowl his fear and disapproval on us, trespassers that we were, and poachers in his front yard.

A week later, travelling this same road, we passed still farther up the brook, to where it winds like the classic Meander through sunny meadows, where the alders and the willows grow, and where the cattle love to come and stand, midleg deep, in its stream. And later still, the photographer of our party tramped to the brook's source, up on the hills, and brought away a picture of its headwaters. But sweetest of all our memories will be that bright morning when we wandered to the brookside, with a little child for company, and lay stretched on a greensward shaded by the meeting boughs of a maple and a butternut, while she played like a baby naiad in the stream, and the brook sang, and the trees whispered, and the birds hopped on branches close beside us, and the kingfisher from down-stream dropped in to call, and the tenant frog stared at us from his pool, and the oxen in the next lot sent looks of fellowship across the stone wall, and we seemed to blend our lives with that of the brook, and for each of us, child, man, and woman, the poet's word was true :

> "Beauty through my senses stole ;
> I yielded myself to the perfect whole."

I suppose all this prattle about the brook seems very aimless and short of the mark to my friends the fishermen ; and they will doubtless feel a wondering disdain for a man who can waste time on a brook without a rod and line. Yet I think I have fished well in New England's brooks, and have brought home as much as they, though I never cast a fly or killed a trout.

> "Hast thou named all the birds without a gun ?
> Loved the wood-rose and left it on its stalk ?
> .   .   .   .   .   .
> Oh, be my friend, and teach me to be thine."

19. THE GREAT CLOUD DRIVE.

No clouds at dawn, but as the sun climbed higher,
   White columns, thunderous, splendid, up the sky
Floated and stood, heaped in his steady fire,
   A stately company.

   .   .   .   .

And brought the rain, sweeping o'er land and sea.
   And then was tumult ! Lightning sharp and keen,
Thunder, wind, rain,—a mighty jubilee
   The earth and heaven between !

CELIA THAXTER.

THE GREAT CLOUD DRIVE.

IN the New England lumber country the spring freshets bring down the harvest of logs which the wood-choppers have reaped with their axes. I judge it to be a most impressive and exhilarating sight when the streams are choked, and the waters fight and foam to make their way, and the rivers run almost solid with the crowding tilth of the forests. I never saw the spectacle. But somehow the term by which it is called—a "drive" of logs,—has been running in my head all this summer, as I have watched the great cloud drive which has filled the air, and choked the channels of the firmament, and surged through space for one whole month and more by the calendar, till we who looked grew dizzy, and the eye fairly ached with the movement of the swift and shifting procession.

The drive began with the great rain that filled all the streams, and caused a midsummer freshet, and flooded acres on acres of tilled land and mowing-fields. As the storm came on, the sky filled with leaden-grey clouds, through whose lower tiers one could see a second and a third layer, forming what John Burroughs so neatly characterises as "three-ply" clouds, which always mean foul business. For two days they hurried by in solid masses which never broke and scarcely thinned out. And all the time, hour after hour, in monotonous persistency, the pouring rain hurled its drench upon the soaking earth, till the mowing-fields were lakes, and the brooks were rivers, and the rivers grew to Amazons and Mississippis.

On the third morning the wind had swung into the north-west. The clear blue sky gleamed through the clouds, which had thinned out into a single thickness of woolly cumulus. The cool wind kept the clouds apart, the channels of the upper deep were fairly clear, and we could see and enjoy each particular cloud, as it hurried past, creamy white in its grand central masses, fraying out all the time into mere ravellings of vapour, through which the blue on which it floated appeared in ever-growing expanses. But toward evening the wind subsided, and the strong sunbeams drew up more vapour which the chill of the upper air turned into more clouds, and this new supply came down-stream with the wind, and the presence of the twilight chill squeezed new showers out of them. Thus the day closed with most brilliant effects,—squalls of rain on distant hills, contrasting with a serene gold and crimson sunset; a dark wind-cloud over the mountain to the southwest, and pink-topped domes of cumulus in the south; finally, deep-grey curtaining clouds all over the heavens, and an evening without a star.

The Great Cloud Drive, Egremont.
(Across Hills toward Mount Washington.)
"*A single thickness of woolly cumulus.*"

which always mean foul business. For two days they hurried by in solid masses which never broke and scarcely thinned out. And all the time, hour after hour, in monotonous persistency, the pouring rain forced its drench upon the soaking earth, till the mowing-fields were lakes, and the brooks were rivers, and the rivers grew to Amazons and Mississippis.

On the third morning the wind had swung into the north-west. The clear blue sky gleamed through the clouds, which had thinned out into a single thickness of woolly cumulus. The cool wind kept the clouds apart; the channels of the upper deep were fairly clear, and we could see and enjoy each particular cloud, as it hurried past, creamy white in its grand central masses, fraying out all the time into mere ravellings of vapour, through which the blue on which it floated appeared in ever-growing expanses. But toward evening the wind subsided, and the strong sunbeams drew up more vapour which the chill of the upper air turned into more clouds, and this new supply came down-stream with the wind, and the presence of the twilight chill squeezed new showers out of them. Thus the day closed with most brilliant effects,—squalls of rain on distant hills, contrasting with a serene gold and crimson sunset; a dark wind-cloud over the mountain to the south-west, and pink-topped domes of cumulus in the south; finally, deep-grey curtaining clouds all over the heavens, and an evening without a star.

Then came a series of days in which the great drive went on with endless variety but no cessation. The mornings dawned fair and promising, and it would seem as if the last of the procession had drifted by us in the night. But the middle of the forenoon would see their tops gleaming again in the sun, as they came rolling and tumbling before a fresh breeze or drifting swiftly before a lighter air. Then, by noon, the smaller detachments massed themselves, darkened into nimbus beneath, dropped a few bolts of lightning and a torrent of rain, and then littered and obstructed the sky till sundown. Sometimes the cumulus would be backed and reinforced by those clouds of the upper regions which foretell a storm; and then the tints would fade out of the lower clouds, they would flatten into a monotony of grey, and another day of rain would make the preparations for more cloud processions next day.

I have often wondered just where the initial point of a thunder-storm might be, and wished that I might see one start. The great cloud drive has afforded plenty of examples. This charging of the air with moisture; this huddling of the clouds together; this liberation of electricity with the changes in temperature,—has created a score of electrical storms before our very eyes. Sometimes they have come in small and weak detachments; sometimes they have advanced across a third of the horizon. Twice I have seen one form almost in the zenith, and start from our own village. Twice, moreover, I have seen a shower make up to the north-east of us, and take the wholly erratic and unusual course of moving south-west, and almost west. In all these cases, the nucleus of the storm was a mass of cumulus which seemed to draw all the neighbouring clouds into the warm vortex of its own ascending vapours, so that the cloud grew by attraction, spread over a continually larger area, and, acquiring intensity and motion with its growth, moved off at last, a full-fledged storm, to deal dampness and devastation far and wide. And, in its development, the clouds seemed almost to reach backward, to roll up in the rear of their actual direction, and thus create a sort of an eddy which caught whatever smaller clouds were moving near and swept them into the main current.

Speaking of currents, I am reminded that nothing about the great cloud drive has been more interesting than to watch how it has revealed certain tracks in the atmosphere, through which the clouds seem to drift as naturally as waters between river-banks. Are there channels in the air, in which the clouds run by some atmospheric gravitation ? They certainly follow regular routes, and their course can be predicted as accurately as the flow of the streams on a watershed, or the currents in Long Island Sound. Clouds may form anywhere in the heavens and move in any direction in which the wind lists to bear them. But just as soon as they begin to flow together, like the trickling rills that run to the brooks and the rivers, they seem to seek certain set direc-

The Great Cloud Drive.

tions; and the local weather-prophets will tell you beforehand when the shower is "going around," and when "we are going to catch it."

There is a range of mountains south-west of us along whose line there is one great channel for the shower-clouds; and those that form to the north take an entirely different trend. There seems to be a sort of cloud-shed which divides these two systems, and turns part of them down into Connecticut and another part eastward toward Boston. But the other day, the clouds seemed to strike some sort of a snag which scattered and utterly dispersed one or two very portentous-looking squalls. A cloud no bigger than a man's hand had grown into a dense, dark storm, heavily marked by the lines of falling rain and seamed by vivid lightnings. It came rolling down toward our plain as if it were going to belch wind, rain, and hail upon us; and we all scurried to shelter as fast as we could. But when the wrack reached the line of hills which marks the channel for east-bound showers, it was turned off its course as sharply as if it had struck a jutting wall of rock. One splinter of the storm flew off in our direction, bringing a puff of wind, a dash of raindrops, and one or two feeble lightning flashes. But the bulk of it went helplessly down the regular channel to the east.

There never was another such season for the study of cloud colours. The chromatic effects of the great cloud drive have been simply endless in variety, swiftness of change, richness, and intensity. There has not been a cloudless sunset for more than a month ; and that implies a wealth of colour effects at that attractive hour. But crimson and gold are not the only nor the most attractive colours in the scale. Such marvels in greys and whites and blues were never wrought into the lower firmament before. Such pinks and bronzes never tinged the great thunder-heads to the east and south, in afternoon and sunset lights. Such contrasts between the angry blue-blacks of the rain- and wind-clouds with the clear, deep azure of the sky behind them, never bewildered the marvelling eye. It is useless to attempt any transcription into words of what cannot be conveyed except to the eye, in colour itself. But the lover of clouds will count it one of the compensations of this soaking summer in New England, that it has been the occasion of the most splendid display of cloud-forms and colours for many a long year.

20. THE FERNS OF THE WOOD.

As oft the pictured scene upon the wall
Brings back to mind scenes dearer and more fair,
As, heard at night, some simple plaintive air
Awakes a chord we thought beyond recall ;
So do ye bring, O dainty, feathery ferns,
The summer's vanished glory to my room.

EMILY S. FORMAN.

THE FERNS OF THE WOOD.

A FEW years ago I was like the vast majority of mankind, who do not know one fern from another, and to whom the common brake is all one with the graceful dicksonia. Looking at their waving fronds as they were crowded in the forest glades was like looking over a strange audience of men and women,—they were only a great mass of verdure, where the eye made no distinction of individuals.

But one day, I was sauntering along the roads of Tamworth with a charming woman and accomplished botanist. She made frequent pauses by the way and numerous détours into the woods; and from one of her excursions she returned with a little fern which she told me was a *Phegopteris hexagonoptera*. The name staggered me, and added to my respect, already large enough, for my companion, which was hardly abated by her condescension in telling me that the plain English title was beech-fern. I picked up a few more names from this good friend of mine and learned to recognise a few of these wayside people by sight. But I had the mistaken notion that ferns are very hard to understand and that it is reserved for expert botanists, and for them alone, to be on speaking terms with what they love to call the "filices." The thirst for knowledge, however, was on me, and when, a few years later, on these Berkshire roadsides I found the ferns more luxuriant than ever, and in larger variety, the determination seized me to know something about them, and, on the foundation laid for me by the wise woman of Tamworth, I went to work to build a little larger information. And it is for the encouragement of those who still stand outside this delightful circle of acquaintance and despair of ever gaining an introduction, that I record this personal experience, and some suggestions for the benefit of those who love the ferns, but know them not.

They are not hard to become acquainted with. If one begins at the beginning, and works patiently, a few weeks will give him a fair acquaintance with the common ferns. A good magnifying-glass is all the equipment which is absolutely necessary; and *Gray's Manual* is a good enough guide to start with. Then, if one knows one or two ferns to begin upon, let him take them as examples and enter upon his course of study.

The key to the identity of the fern is the spore-case upon the under side of the frond. Here are ripened the spores which, falling into the ground, mature into the true seeds of the plant; and the ferns are classified according to the shape of these "sori," as they are called; and the way in which they are covered, and their arrangement upon the divisions of the frond distinguish the varieties from one another. Some of them are round, some kidney-shaped, some oblong. In some the covering which protects them is caught down in the centre, and opens at the edges; in others it is caught at one edge and opens along the other; in others it is attached underneath and opens at the top. In the dicksonia the covering is a corner of the leaf turned back upon the seed-dot. In the common brake the whole edge of the leaf is folded over, like a hem on a piece of cloth. A few ferns, like the sensitive-fern and the royal, devote special fronds to the task of producing the spores and are easily made out when the time comes for them to ripen. But the point is that the fern student has, first of all, to examine those little excrescences on the under side of the leaf, which so many imagine are some foreign and hurtful growth upon the plant; and it will not be so very long before he will be able to make out the differences very handily, and name his specimen with accuracy.

Of course this is not the only means of identification. The root-stock is important, and so is the frond, with its divisions and subdivisions. The veins of the leaves play a conspicuous part, and very soon become almost the only feature required to distinguish the individual. When that time comes to the student every wood-walk and every stroll along the highway becomes a social meeting. For everywhere one encounters his acquaintances and friends, and they seem to nod recognition as he passes and to beckon friendly greeting.

Not only are the characteristics of the ferns comparatively easy to master, but the family itself and its relations are not numerous. The varieties found in New England and the Middle States are soon learned; and there is a fine sense of mastery in knowing pretty thoroughly the appearance, habits, and haunts of one whole family of the plant world. For one, I can say that hardly any knowledge that I ever acquired has given me more delight than my imperfect and desultory acquaintance with the ferns.

In the matter of guides there are now a number of good ones to choose from. *Gray's Manual* is a little out of date, but it is clear and helpful. *Our Native Ferns and their Allies* by Prof. Underwood is a capital work, and there is a compact and cheap booklet by Raynal Dodge called *Ferns and Fern Allies of New England*. The new *Illustrated Flora of the Northern United States and Canada* by Britton and Brown contains a most admirable section upon the ferns and allied forms. And of course, if one is lucky enough to have access to them, the magnificent two volumes of Eaton's *Ferns of North America* are beyond anything else in their beautiful coloured illustrations and their minute and accurate descriptions.

Now I hope I have convinced the reader who has cared to follow thus far that it is not a hard thing to know the ferns; and I trust that he is ready to buy him a magnifying-glass or a botanical microscope and to go forth in search of new acquaintances. I only wish that I could take him for a stroll through my own fernery. It is rather larger than the books recommend, larger than the millionaires build, who import their ferns from Brazil and from Abyssinia. For it is about a mile square, in its main departments, with several annexes, of smaller dimensions, twenty or thirty rods each way, perhaps; and there is one very interesting section, where my polypodies grow, which is about twenty-five hundred feet in the air. In this fernery I have constantly growing, in the season, the most interesting varieties of New England ferns. I take the greatest pleasure in studying them from day to-day, as they uncurl in the spring, and glow with their rich greens in midsummer, and turn pale in decay in the autumn days. I feel an honest pride in taking the few friends who really appreciate them on the tour of this splendid fern garden. And if it be argued that I have undertaken to carry on the culture of ferns on too elaborate and expensive a scale for a man on a limited income,—a dominie at that,—I answer that I have so far got out of my enterprise without its costing me any more than the shoe-leather which I have worn out in looking my possessions over, and in watching the growth of my pets.

The first room to which I am fond of conducting visitors is a delightful knoll, a score of rods from the house, whose ledges are covered with a grove of noble pines, and hedged about with the common shrubberies and vines which love such rocky nooks. On the northerly exposure, just under the jutting rocks, is a group of the dainty *Cystopteris,* a frail and delicate plant, with curious bulblets growing on the under sides of its fronds, which fall away and propagate the plant. Close beside these we find little clusters of the ebony spleenwort, whose polished stems gleam between their dark-green pinnæ. Each in its own way, these are among the very daintiest in the collection, though most people pass them by unnoticed. Down in the swampy grass a rod or two away from them, I sometimes raise one of the small *Botrychia* or grape-ferns, another tiny slip which would easily be mistaken for some flowering plant. And if you care for these little things, and will step across the field, climb a rail fence, and go up the hillside to where the limestone ledges crop out, I will show you in some crevices of the rock two or three places where, every year, I am sure to find a little colony of the small spleenwort, the *Asplenium parvulum,* perhaps the tiniest specimen in all my collections.

But the largest room in my fernery and the one which furnishes the largest variety of specimens, is the wood-lot yonder, within whose territory, half a mile square, one may find nearly all the common wood and swamp ferns of New England. At the very entrance to the path the ground is grown thick with the pale-green fronds of the New York fern, whose grace and beauty are not one whit diminished because it is so common ; and before many steps are taken the rich hay-like odours of the dicksonia delight the sense. Here the sensitive fern, the *Onoclea sensibilis,* spreads its hands upward, though it clings pretty closely to the earth, and the strong, masculine *Pteris aquilina,* the common brake, rears itself in an assertive sort of way above the rest of the ferns. There is great abundance of *Osmundæ* in this wood, all three varieties being well represented in the collection. One soon learns to know the *Osmunda claytoniana* by its brown fertile leaves, interrupting the pale green of its great fronds, and the *Cinnamomea* by its seed-leaves growing within the centre of a little circle of sterile fronds.

Here, too, there is much of the royal fern, the *Osmunda regalis,* whose title to the throne some fern lovers would decidedly dispute. When I was making my first essays in the lore of the ferns I had a very curious experience with this one; I was watching everywhere for the royal fern, which the books said was very common; and yet I was able to find nothing which seemed to answer the description given in any of my authorities. At the same time I was immensely puzzled by a bush which I met everywhere, and which bore a fertile leaf that was extremely fern-like. At last in my ignorance I sent a specimen to a wise woman, who in turn sent it to a wiser woman, who returned the word that the fern I did not recognise was the fern I had failed to find, and that my search for his majesty, the royal fern, would have been shortened if I had more carefully studied the unknown variety which had baffled my small knowledge. They were one and the same.

If one is proof against mosquitoes and fearless of

wet feet, there is a corner of this part of the fernery where some fine examples are to be found of a number of the choicest varieties in the collection. It is less than a hundred yards, straight in from the road, to the heart of the swamp which occupies most of these woods. The thicket is filled with that many-named frond which the latest books call the Christmas-fern, alias sword-, shield-, black-, and rock-fern ; one may take his choice of names,—the fern remains the same elegant, dark, glossy evergreen, which florists put in all the church bouquets, and which holds its own against all vicissitudes of weather and season. When we reach the bog, we are in the midst of a profusion of ferns that is almost tropical. Clayton's fern, the cinnamon, the royal, grow rank and tall ; dicksonias crowd thick and fragrant ; and the sensitive-fern fairly carpets the ooze. The silvery spleenwort makes its home here, and Boott's shield-fern, with reddish-gold seed-spots decorating its maturing fronds. But most beautiful of all, a picture of waving grace, grows the glorious ostrich-fern, a circle of out-curving fronds, each one of them as perfect as an ostrich-feather, and all together making a green vase or urn, as beautiful among ferns as the elm among the trees of the meadow. Had I the christening of these varieties, I would crown this one with the royal title, for its regal dignity and stateliness.

One could linger here for hours in contented study of these teeming forms,—if it were not for the mosquitoes ; but these swarming pests make even a few

The Haunt of the Ferns.
(In Woods near Hubbard's Brook.)
"This vast fernery of all outdoors."

UNIV. OF
CALIFORNIA

moments' stay an act of heroism. The best one can do is to fill one's hands full of fronds, and run for dear life. But the spoil is well worth the venture.

If I could count upon the patience of the reader, I should love dearly to tell of the annexes to the fernery; one of them under a rocky bank, at the other end of this swampy wood, where the walking-fern creeps and roots itself in company with the tiny spleenwort; another on the hillside, half a mile farther on, where the maidenhair still grows abundantly, having escaped the plundering hand of the spoiler; or of that mountain forest, four miles off, as the crow flies, on whose crags and ledges the deep shades of the evergreen wood-fern, *Dryopteris marginale,* draw the eye and the hand of the climber, and where as soon as the ascent begins the common polypody gladdens the sight with its clean-cut outlines. These I count as parts of my fernery, rather remotely placed indeed, but all the better adapted by their situation for the best growth of the varieties to which they are set apart.

There is one satisfactory trait about ferns, and that is their constancy. They cling to their old haunts. One finds them year after year in the same spot, true to the patches of soil where their ancestors grew, ready to welcome again and again the feet which have learned the way thither. It is partly to this "homing" instinct of the ferns that I owe my fernery, with its variety of forms ample for study and enjoyment compressed into so small a territory that a half-hour's walk will enable one to gather a score of different kinds. Nor is there a square mile of New England countryside where almost the same variety may not be found, to give the patient student a love of these charming plants and help him to know their native haunts. For one, I have to thank that wise gardener, Dame Nature, for the fernery she maintains for me, free of expense, as long as I care to use it. I should not dare to trust my own skill in constructing and keeping up a fernery; for I am no gardener, and it is sure death to any growing thing to attempt to cultivate it in my house. But I can always trust the skilled Grower of green things whose care keeps up this vast fernery of all outdoors in such attractive shape for me. And the chief glory of my fernery is that it is everybody's else as well.

21. LIVING WITH A LAKE.

I know no gladder dreaming
In all the haunts of men,
I know no silent seeming
Like to your shore and fen ;
No world of restful beauty like your world
Of curvèd shores and waters,
In sunlight vapours furled.
W. W. Campbell.

LIVING WITH A LAKE.

ONE of the necessities of a complete life is to have had intimate and varied relations with water in all its scenic forms. Man can never feel that he knows nature until he has dwelt near to every natural division of water long enough to become thoroughly sensible of its influence, its changing phases, its distinctive effect. One need not subscribe to the old saying of the nature-lover, "Water is best." But one must at least know how good water really is, in its ministry through the eye to the spirit, ere he can be said to have passed even the novitiate's degree in nature lore.

I have had all the joy and enlargement which come from intercourse, if one may so speak, with all the chief forms of aquatic scenery and surroundings. I have known the charm of rivers; and pictures of the industrious Merrimac and of the leisurely Mystic and of the classic Charles still rise to gladden many a retrospective thought. I have rich recollections of the sea, in every aspect, under every sort of a sky, beating or rippling upon all sorts of coasts. Countless brooks of blessed memory make music in my soul to-day, as sweet as when I heard with ear of sense the gurgle of their waters in the forest, or over the stones in open meadow-reaches. I have seen, with homesick heart, how the great lakes mimic the great ocean, which they can never counterfeit with success to the eye trained to look upon the sea.

But a little corner of experience has been lacking, to make it all complete. It has still remained to try the pleasures of a few weeks' life by the side of a lake; not a vast inland sea, nor one whose opposite shores always wear the haze of distance,—but a small, a convenient, a manageable lake, not too large to grasp and comprehend in all its aspects nor to allow the feeling of intimacy and even spiritual ownership. Such an opportunity has offered at length; and it becomes a sort of duty to record one's impressions.

M. Michelet has a theory, which he presents in a very interesting book, that mountains are a sort of organism; that they grow, and decay, and pass through stages analogous to those of organic life. Doubtless most of us have felt the power of this thought when we have stood face to face with some grand peak, or have watched it from day to day. But whether M. Michelet, or the average observer of nature, would credit a lake with any such approach of personality, I hardly dare affirm. There seems, however, to be a trace of this distinctive character, a sort of low form of vitality, about a lake. It has its moods. It impresses its individual life upon one. It comes to stand for a certain sort of companionship, much like that of the mountain. It is less masterful

Lake Pontoosuc, Pittsfield.

(Mount Greylock Range in the distance.)

"*A manageable lake, not too large to grasp and comprehend.*"

the corner of .... has been licking
I all complete. It has still remained to try
...... a few words ... by the side of a lake;
.... opposite shores
... ... of ....,—but a small, a
... . lake, not too large to grasp
.... to allow the
.... individual ownership,
.... at length; and a
.... one's impressions.
...., which he presents in a
...., that mountains are a sort of
.... they grow, and decay, and pass
.... to those of organic life.
.... of us have felt the power of this
...... we have stood face to face with some
...... have watched it from day to day. But
M. Michelet, or the average observer of
..... credit a lake with any such approach
...., I hardly dare affirm. There seems,
to be a trace of this distinctive character,
low form of vitality, about a lake. It has
...... It impresses its individual life upon one
.. stand for a certain sort of companionship,
that of the mountain. It is less masterful

than the mountain. But for that very reason perhaps, it is more companionable. It does not impress its own mood upon the spirit of man. It rather lends itself to his temper, and blends with his humour. Stability is the word which defines the soul of the mountain; flexibility is the characteristic of the lake.

Chief of all its charms is an infinite play of lights and shades, hues and shadows, colour in constant flux, subtle blendings of tints in all keys and tones, incessant alliances with shore and sky whose issues are rich in all the resources of colour which can be conferred by earth and sky and water itself. The lake has its own individual key-note of colour, which it takes from the rich greens of its wooded shores, mixed to a darker shade in the depths of its own waters. But in that particular key the variations are almost endless. Not a cloud drifting across the sky; not a breeze rippling the water's surface; not a change in the angle at which the sunlight falls; not a variation in the humidity of the air; not a shift of colour in the foliage in spring or autumn days,—that does not find an instant report and correspondence in the upturned waters. There is no more delicate colour-gauge in all nature. The mountains are responsive to the same influences, but they do not begin to be as subtle, as sensitive, as vivid in the report they render to the eye.

When you watch the sky as it is reflected on the mountain, you think of the mountain; when you watch it in the lake you think of the sky. But every instant of the day, and even in the darkness of the night, the lake offers to the eye some new combination, some surprise in colour, some attractive disposition of light and shade. Last night it was so grey and monotonous under the high fog which drifted in with the twilight that it could have been reproduced in monochrome. The night before under the sunset light that flushed, not the west alone, but every inch of space in the firmament, it was a weltering tide of subtlest pink and rose. This morning, reflecting the steely clouds which have slowly spread over the heavens, it has all the neutral sheen of a mirror. To-morrow, when the north wind blows, and a clear blue shines in the sky, its rushing waves will darkle into indigo and cobalt, picked out with the flecks of the foam. And so from day to day and from hour to hour the lover of colour who lives with a lake has incessant joys as his portion. His eye is gladdened with a chromatic play which never grows wearisome, never satiates, an endless optical symphony.

But man cannot live by colour alone. Nor is that all which the friendly lake offers to beguile the days. Its sinuous shores invite the feet of the stroller, and the skiff of the loitering rower. Within the compass of its ten miles of shore line it offers the large variety of a little world. Its head waters reach the banks of long upland meadows, stretching away to distant ranges, while it finds its outlet toward the sea in a notch between steep, wooded hills which have all the semblance of the Adirondacks. Here its banks are dressed in hemlock, maple, and chestnut, and there the bright green of the willows at the low water edge contrasts with the darker tints of the elms in the meadows behind. On one side, bold, rocky ledges fall abruptly away into deep water, and opposite, the sedges and the rushes grow far out into the shallows. This is the rich green summer setting of the lake. How fortunate the eye privileged to follow all the transformations in the foliage and herbage around it, from the early, tender greens of April, till the last brown leaf is whisked into its waves by November gales!

Companionship with the lake involves, of course, some degree of association with the forms of life which surround it, of which it is a sort of centre and rallying-point. This includes, perhaps, the human beings who haunt its surface, and linger about its shores. Though really these are rather adjuncts, foreign to its real life, than part of its being, or essentials of its character. They seem like the flies which crawl over the body of a man, yet are no part of him; and so far as the lake is a spectacle, it cannot be said that its human neighbours succeed in making themselves very distinguished or important as elements thereof. They go pulling about over its bright waves in their little skiffs; but the glint of their oars is only one more flash added to the millions which greet the eye from every wavelet and billow. Or they sit dejectedly hour after hour, watching the floats of their fishing-lines, which rarely give any sign of being agitated by the wary fish below; and in this relation to the scene they are less interesting than so many floating logs. Only when they plunge into its depths, and as bathers and swimmers become a part of it, do they float up, so to speak, to the level of its own charm upon the eye and the mind.

But it is far different with the other creatures who inhabit its shores, and soar or flit above its waters, and swim within them. These belong to the lake. Its life and theirs blend as the sky and the clouds mingle with the waves. One thinks of the two together, as they belong. The birds love its banks, and frequent them as an evidently approved summer-resort. Nor is there any doubt in my mind that in the days and weeks of the mild southern winter, there is much chatter and chirp in the pine groves and the rice-swamps over the charms of this far northern lake as a place where it is good to rear one's brood and get cheap and wholesome fare, during the summer months, with small liability to gunshot fatalities. The swallows hereabouts are plump and lively, and when they skim and dart and dive after their suppers in the glow of the sunset, it is evident that they are faring well and taking far better catches than the stolid fishermen around whom they circle.

A little later, down among the trees on the point where the willows grow, there begins a tremendous chatter, as these lively little fellows go to roost, a clear case of easy, satisfied, self-sufficient gossip, full of importance in the bird-world, and amusing even to the dull mortals who can only half understand it all. In the fields which skirt the water's edge the robins love to forage; and one day I had the pleasure of assisting, as a spectator, at the first flight of a brood of nestlings over which not only the whole robin family, but half a dozen other bird households watched with noisy solicitude. There is a family of belted kingfishers living down by the mouth of the brook yonder which never fails to protest against our invasions of its premises, in a sharp, rattling note like the drawing of a stick along the fence-paling. Nor do we ever lose an opportunity to beat up the handsome pair and set them to clattering and plunging from one thicket to another; for they are beautiful to behold and in perfect harmony with their surroundings.

One day as we were floating idly near the sedges which grace the mouth of the brook, suddenly there came dropping down from the air the gaunt form of a bird whose long bill and lean breast were matched by the slim legs which he pulled behind him. With much awkward flapping of his wings he settled among the reeds and grasses; then we knew that we were honoured by the visit of a heron. Breathlessly we awaited his movements, and hoped he would not find our presence objectionable, and that he would deign to trust in our good will. But the diffidence and shyness of his retiring nature could not be overcome by any telepaphy of ours; and presently he awkwardly shifted still farther into the thicket; and then seeming to make up his mind that where there were so many inquisitive spectators was no place for him, he floundered into the air again and was soon sailing calmly to some quieter retreat. A few days later he reappeared with a mate; but this time did not deign to alight, ignoring our attractive marsh, possibly on account of its remembered publicity.

But the chief lion in our bird community,—if one may be pardoned for putting it thus,—is an eagle. He is a splendid fellow, and it is the episode of a day when he appears in the sky spaces, soaring with that stately ease which so mocks our human resources of motion, or rushing forward upon his way. He seems to have learned the immunity he enjoys under the laws of the State; for he is a fearless creature, so far as man is in question, and permits his human admirers to come within easy speaking distance of him when he is resting upon some favourite perch. How the smaller creatures regard his presence I am not prepared to say, nor whether the king-bird who bullies the crows and the hawks, ever ventures to chase an eagle! But for those of us who have nothing to fear from him he is one of the joys of life upon this water-margin.

The lake has fallen a little from its high-water level, and there are rusty patches on the wooded hillsides which suggest the waning of the summer. The lake and its verdant setting will soon be entering upon another phase of their life. The bullfrog seldom sings any more to his lady love; but the katydid sharply reiterates at night his absurd insistence—"she did," "she did." By these and other infallible signs we know that the friendly birds will soon vacate their summer homes. The stripping of the trees will shortly begin, and before we realise that it is time, there will be a hoar frost in the swampy field yonder, and the sharp needles of ice will dart to and fro on the black water, and some fine morning there will be six inches of hard, cold, glassy tiling laid down over its surface.

But long before that, our life with the lake will be only a sweet dream of the summer, a memory to be called up in winter days, and to be grateful for always.

22. THE FRUITAGE OF BEAUTY.

"Happy," I said, "whose home is here !
Fair fortunes to the mountaineer !
Boon Nature to his poorest shed
Has royal pleasure-grounds outspread."
R. W. Emerson.

THE FRUITAGE OF BEAUTY.

The scene: A maple-shaded lawn looking out upon the encircling hills of Berkshire.

The time: Sunset and the few minutes just after, in late August.

The people: The Dominie; The Wife; Lisbeth; Adelaide.

The Dominie. Well, the summer is over and gone. That rack of clouds in the west is decidedly Octoberish. We have come to the harvest weeks; and it seems no longer ago than yesterday that I was gathering the marsh-marigold by the brook and the hepatica in these woods. How quickly the season has gone, and all the glory of it!

The Wife. Ah, but has the glory of it gone? Can it ever go? Once seen must it not always live in your mind, a memory at least, and a harvested joy of life?

The Dominie. That, of course, depends. If I have farmed well in the beauty of these scenes, I shall carry away from them a rich freight of harvested impressions. That I grant without question. But how many, do you suppose, of all the thousands who have been living in the midst of these beauties, will take away any distinct impression, any clear, strong memories of the things they have seen?

Adelaide. You might carry your question farther and ask how many of these people have really looked at these things enough to have seen them at all. And for those who do not see at all, as well as for those who see only to forget, the glory and the beauty are indeed quite passed away.

The Wife. Is it as bad as that? You make the case quite hopeless for those who have but just begun to observe, who see but little, yet who are growing in perception of nature and her beauties. They may get but light impressions; but are these necessarily void? May they not be the beginnings of a deeper and more adequate appreciation of all this beauty?

Adelaide. I am afraid I have not much hope for those who are not endowed in the beginning with some love of the beautiful in nature. No eyes, no sight. No soul for beauty, no apprehension of it. Think of all the people who have lived here for years, yet who have no more sense of all this glorious scenery, this light, this colour, this sky-prospect, than the house-dog or the cows.

The Dominie. But do you think they have no capacity for beauty, or that they and their children might not be led to a larger enjoyment of it, if only they had a proper impulse and direction? I should be sorry to believe that there is anybody who is

Williamstown Hls, Williamstown.
(Looking tow'd North Adams.)
"The encircling hills of Berkshire."

CRITTOGRAMMI DI VITA

The Fruitage of Beauty.

utterly without this sense of the beautiful, at least in some germinal form and degree.

Adelaide. Well, I doubt whether it is worth one's while to try and develop what, in some folks at least, would prove to be very poor seed, in very bad soil. The time and trouble were better put in somewhere and somehow else.

Lisbeth. Yes, but suppose you had children whom you wanted to teach the love of all beauty, and this kind in particular, and found them slow and dull and indifferent, would you stop trying to impart to them what they lacked? Because they were born deficient would you let them grow up and die in the same lack? Would you not try to round them out on their weak sides?

Adelaide. I never would try to make a silk purse out of a sow's ear.

Lisbeth. That is all right. But did you never hear of educated pigs; and were you never struck with the wonderful things that even a pig can be taught, if only his trainer has patience and takes the necessary time?

The Dominie. Let me stick a pin in there. I am quite sure that Lisbeth has used a most vital word. The time element has very much to do with the acquisition of this love of nature and its beauty. Such scenes as this must have time to soak into both sense and soul. An appreciative lover of nature is not made in a day nor a season. One must come again and again to such beauties in order to absorb them, realise them, carry them away in memory. People who have the right to speak on the subject say that it is necessary to live for weeks in the presence of Mount Washington or the Matterhorn, before one can feel that he begins to understand them.

The Wife. What you say is borne out, too, by the lives of those who live long in the midst of such scenery. It must be that it makes its impression. Else why is it that when the mountaineer goes away from his hills he misses them, longs for them so, loves to go back to them. The same is true of the dwellers by the sea or on the plain. It must be that they unconsciously learn to love these grandeurs and these beauties. Long years spent among them serve to impress them deeply on the spirit. They enter into the life. I am not sure, either, that the people who live among these beauties are insensible to them. Why do we say that?

Lisbeth. Isn't it because they never say anything about them, in appreciation or in praise?

The Wife. That is just it. But impression and expression are very different things; and a good many of us, I suspect, feel a great deal more than we do or can tell. I do not care whether my children grow up with the power to talk finely about nature or not. But I do want them to feel finely its noble influences.

Adelaide. Yes, for nature is the great refiner; and the love of beauty in good hands ought to impart a grace and a delicacy to any life. We all need it and ought to have it ; and of course I would be the first to admit that they who are altogether or largely deficient in aptness for this enjoyment and culture are great losers in the experience of life.

The Dominie. I am inclined to think that they lose more than mere enjoyment. I have grown to feel that the love of nature and its beauty and inner life have much to do with the enrichment of the religious life. Religion has been the gainer both from science and art, for these interpreters of nature have broadened our vision, lifted our ideals, and expanded all our conceptions of the universe and of its Creator.

Lisbeth. But people can be religious without having this love of nature and its beauties. Did the Puritans care much for nature ?

The Wife. And were not men devout and even nobly inspired in religious feeling, long before they knew much science or art ? What did David owe to the doctrine of evolution, or to Corot or Constable ?

The Dominie. Not too fast, please, either with questions or conclusions ! I only spoke of the enlargement of religious ideas through contact with nature. The Puritan would have been a more religious man if he had been more susceptible to these beauties of the outward. And as for David——

Adelaide. Now don't try to tell us that we have outgrown the Psalms, and that Spencer's *First Principles* is grander than the Nineteenth Psalm. I will not listen to any such nonsense.

The Dominie. Well, I do not ask you to. I am no such heretic toward the ancient landmarks as you try to force me to be. But suppose David were to come back here now, and, instead of the hills to which he lifted up his eyes, see Tacoma, and Sir Donald, and Jungfrau; and suppose instead of the shepherd's simple knowledge he were to hear the heavens declare the glory of God with the ear and understanding of Herschel. What might we not hear in the devout lyrics he would sing for us?

The Wife. I should hope they would not lose anything of the simplicity and childlikeness of the old ones. The world would be the poorer if they did.

The Dominie. True enough! The world has not lost and will not lose the primitive and germinal faiths which were born with it. But I devoutly believe that its culture and knowledge are all the time expanding these childlike ideals, and that the same spirit, with the maturity of the centuries added unto it, would strike a grander note.

Adelaide. Well, I am glad you think properly of the ancient harvests men reaped in this field of beauty.

The Dominie. Good! and I hope you realise that the soil grows richer with the deposit of the ages.

Lisbeth. Hark! I hear my baby cry.

The Wife. Well, it is too damp for us any longer. Let us go in.

23. A QUEST FOR WINTER.

While Mother Nature comes in love to throw,
O'er all, the soft, white comfort of the snow.
                              EMILY SHAW FORMAN.

A QUEST FOR WINTER.

THE winter so far had proved to be an unusually severe summer. It was not exactly a case of "Winter lingering in the lap of Spring," but of September holding over into the New Year. Everybody was tired of the weakness of the temperature, lingering up among the forties and fifties. There was no market at all for sleighs, for the snow-line seemed to have retreated to the north of the Adirondacks and the Green Mountains. Yet I had faith that somewhere, away from the coast, there was winter weather, and in spite of discouraging advices, I started for the quiet Berkshire town which had been a haven of rest in summer days. All through Connecticut, along the shores of the Sound, up the valley of the Housatonic, nothing appeared to cheer my drooping hopes. Everywhere there was the same bare, brown landscape, everywhere were the same stark, leafless trees, the same open ponds, just fringed with ice about the edges.

The train rolled northward into the dusk and the dark, and still there were no signs of snow. Just across the Massachusetts line there appeared to be a faint sprinkling of white, as if a light snowfall had dusted the earth with a tantalising hint of winter. But when Great Barrington was reached it was no sleigh with its tinkle of bells that awaited me, but a comfortable buckboard, suggestive again of summer rides among those Berkshire hills. There was however a most winterish chill in the crisp air, and the January constellations blazed through the dark as they shine only when the mercury is falling well toward zero. The glow of the warm, cheery parsonage which opened its hospitable doors to the traveller imparted a thrill which belonged to midwinter indeed. And when, after the hours of chat which dominies off duty so dearly love, I went to my room, it was clear that the frost was thickening on the white and glistening panes, and the roar of the fire in the "air-tight" stove sunk into a sighing wheeze that lulled me into a dream of icicles and snowdrifts. I had found the missing winter.

The sun next morning was soon overmatched by the clouds which worked up from the south-west, and by afternoon the signs in the heavens above and the temperature in the earth beneath gave more promise of a snowstorm than anything we had seen for months. By dinner-time the clouds had conquered, and there was almost no sunshine at all. Yet the chance for a ride was not to be foregone, and in a summer waggon, but with winter wraps, we drove over one of the favourite routes which we had last travelled in brilliant August weather.

The landscape was bleak and stern under the thin

snow. The colour was all gone out of it. It was like a sketch in black and white for the completed picture to be done in midsummer. The only suggestion of livelier hues was in the saffrons and yellows of the willows, which did their own brave best to put a cheerful glow into this sombre afternoon. But there was soon another element in the scene which made life and movement everywhere. A swift squall of snow, the skirmish-line perhaps of a heavier fall to come, whirled in upon us from the south and the air was fairly alive with the hurrying masses. It was marvellous to see how soon the mountains changed their aspect. The hard outlines softened, the uncompromising details of clearing and grove, gorge and boulder, valley and climbing slope, were merged and melted in this cloud which fell upon them like an enchanter's mist. In the shadowless depths of the grey spaces the Whitbeck range, which rose just in front of us, might have been two, five, twenty miles away; there was a total loss of the sense of distance. There were no aids to the eye to help it in its judgments.

The clouds grew heavier as we swung about the circle which cuts into North Egremont, and then bore around toward Barrington. The first squall ceased. But as the dusk came on, the sky took on a stormy complexion. The wind was in the east, the air was heavy with a frosty damp,

> "A hard, dull bitterness of cold,
> That checked, mid-vein, the circling race

Of life-blood in the sharpened face,
The coming of the snowstorm told."

The Taconic range had darkened with the clouds until it was hard to recognise in that blue-black, frowning mass, the hospitable and smiling slopes which had invited our steps on so many summer days. Mountains have very quick sympathies. They respond as subtly as the sea to every shadow of a change in the sky and the clouds. With this winter twilight closing in, and heavy, black masses of vapour hovering low about their summits, these sometime cheerful hills looked as dark and gloomy as an abode of blue devils, a haunted realm peopled by the spirits of care and trouble.

The night fulfilled the promise of the afternoon. Quietly but steadily until morning the snow fell, and with the dawn stood six inches deep upon a level, good for many days of sleighing. It was with the elation of one who has discovered the thing he has been long seeking that I climbed into my host's sleigh, and behind the same fine roadsters which had made light work of many a summer mile, went bowling along in a glorious winter outing. It is amazing to think how little winter is known or appreciated by the incorrigible cockney, and how little he realises the variety, the interest, and the healthfulness of outdoor life under the sign of the Crab. To-day, for example, one could find nearly all his large wildflower friends of the summer days, grown brown and old indeed, but plainly to be recognised. The golden-rod and the wild carrot, the vervain and the tansy, all erect their slender stalks, and nod a brave greeting.

The trees stripped of their foilage only show their splendid anatomy the more clearly. They look like wrestlers trained to finest condition, with no superfluous flesh, their contours giving the impression of strength and noble vigour, missed under the draperies of summer leaves. This day the clearing skies showed bright and cold behind the scattering clouds, and the crisp atmosphere recalled Lowell's vivid phrase, "The air you drink is *frappé*, its grosser particles precipitated, and all the dregs of your blood with them." I was loth to see the sun go down and end this satisfying winter day.

But the frosty night gave promise of a clear morrow, and when the rising sun set the snows aflame upon the Dome, and every twig on elm and maple glittered as if crusted with gems, and the crisp snow creaked under the footfalls of the few passers-by, it was hard to keep from singing the Doxology aloud, and impossible not to hum it under one's breath. The mercury kept below fifteen degrees all day. The sun shone clear and the winds were still. The wood-fire in the ample fireplace shed a wholesome glow into the cheery room, and the whole house, the birthplace of my delightful hostess, and the homestead of her father before her, was radiant with the comfort and the peace of a typical New England home. Twice that day it was my privilege to greet a congregation made up entirely of the village-folk and the farmers from the outlying country, and feel that fellowship which is broader than sect and deep as the spirit of the common Master. There were no summer boarders now to help fill the little church, so that the preacher could come close to his hearers, and speak the more familiarly on the blessed themes of service and love and duty. This winter worship in scenes so fair, on a day so ideal, was an experience to be most carefully cherished in the memory. The service sanctified the day and the day was a benediction upon the service.

But the winter holiday was not over. Still another exciting episode was to be added to its list of pleasures. The kindly "master of the horse" at the summer hotel was intent upon hospitable things, and out of his brain was evolved the scheme for a January picnic on the mountains. Every summer visitor to this region is expected to go to the wild glen where the water-brooks plunge down Bashbish Falls into the valley of Green River, on their way to the stately Hudson. To Bashbish we must go on a sleigh-ride. The great, four-seated sleigh was at the door by nine o'clock; nine muffled figures were stowed in its depths; and with lunch baskets and luggage we started. The day was threatening. The Dome was shrouded in a cloud, the wind was east, the sky was leaden. But the ardour of the picnickers was too strong to be cooled by temperature or prognostics of storm, and the expedition moved on its way across Gilder Hollow, up the wild ravine which leads past Sky Farm to the corner township of the Bay State, Mount Washington. It was a thrilling ride, along the side of a deep ravine, on the verge of a gulf full seven hundred feet to the brook-bed below. The road was narrow, and it seemed as though if the horses should swerve we should go tobogganing to the very bottom. But twice we met descending teams, and found room enough to turn out. There was more snow here than in the valley, and the sleighing was excellent. And the woods bare of leaves, the long vistas down the gorge, the sharp contrasts to the summer aspects with which we were so much more familiar, made the upward way seem very short, so that, as we drew up before a Mount Washington farmhouse and greeted its astonished owner, we were quite prepared to vote that for real, racy pleasure, January is the most desirable month in the year for picnics. The lunch was eaten with a keener relish than on summer days, albeit it was the conventional bill-of-fare,—sandwiches and hard-boiled eggs and coffee.

It was soon despatched and we had parted with our host, hardly recovered from the shock of our surprising visit, and were rapidly descending another and wilder ravine to the picturesque Bashbish. Just as we were approaching the falls, and before we had fairly caught sight of the Hudson valley, the grey snowstorm pervaded the air, and shut us into short-range views. We found the lovely cascade all cased in ice, a solid rampart at the base, a thin glaze of frosted windows around its falling stream, with canopies and columns, buttresses and abutments, for a hundred feet up the steep cliffs. The snowstorm whistled a lively obligato to our comment of wonder and delight, the gale roared a deeper bass to the clear soprano of the brook. It was worth a score of languid summer visits, to stand here in the driving snow, deep in the heart of the hills, defying storm and temperature for this glimpse of the architecture of the ice.

But we could not linger in these fastnesses of nature. I must meet the train down in the valley, and hasten back to the work and pleasures of the town. These venturesome friends had twelve miles to drive in the teeth of this strengthening gale to their snug homes on the Berkshire plains. In half an hour I bade them good-bye and started southward. They debated the wisdom of remaining over night in the village, or of making the homeward trip in the storm. With characteristic Berkshire courage they decided to push over the mountain for home ; and after a long, hard battle with sleet and snow and wind and cold, they accomplished their journey without damage. Crusted and sheeted in sleet and fringed with icicles, they descended like a party of frost-spirits straight from the lands beyond the pole. Meantime their whilom guest, upon his homeward journey, saw the snow turn to sleet and the sleet to rain, and in the steaming, slushy streets of New York found himself at twilight wondering whether his forenoon's adventure were not after all a dream. But dreams leave no such substantial memories as I had added to my store of Berkshire days. I had found the missing winter under the Dome of the Taconics.

BOOKS FOR THE COUNTRY

NATURE STUDIES IN BERKSHIRE

By JOHN COLEMAN ADAMS. With 16 illustrations in photogravure from original photographs by ARTHUR SCOTT. Photogravure edition. 8°, gilt top, $4.50. Popular edition. Illustrated. $

"The spirit of the region is very happily caught by the author, who is fond of outdoors, and a sympathetic chronicler of the events of field and woodland. . . . The pictures in the book are very fine indeed. . . . The style of the narrative is clear and unaffected, and the book is one that will not easily be relinquished when once taken in hand. The book is attractive and sumptuous, a credit to the printer's art."—*Chicago Evening Post.*

LANDSCAPE GARDENING

Notes and Suggestions on Lawns and Lawn-Planting, Laying out and Arrangement of Country Places, Large and Small Parks, Cemetery Plots, and Railway-Station Lawns; Deciduous and Evergreen Trees and Shrubs, The Hardy Border, Bedding Plants, Rockwork, etc. By SAMUEL PARSONS, Jr., Ex-Superintendent of Parks, New York City. With nearly 200 illustrations. Large 8°, $3.50.

"Mr. Parsons proves himself a master of his art as a landscape gardener, and this superb book should be studied by all who are concerned in the making of parks in other cities,"—*Philadelphia Bulletin.*

LAWNS AND GARDENS

How to Beautify the Home Lot, the Pleasure Ground, and Garden. By N. JÖNSSON-ROSE, of the Department of Public Parks, New York City. With 172 plans and illustrations. Large 8°, gilt top, $3.50.

"Mr. Jönsson-Rose has prepared a treatise which will prove of genuine value to the large and increasing number of those who take a personal interest in their home grounds. It does not aim above the intelligence or æsthetic sense of the ordinary American citizen who has never given any thought to planting and to whom some of the profounder principles of garden-art make no convincing appeal."—*Garden and Forest.*

ORNAMENTAL SHRUBS

For Garden, Lawn, and Park Planting. With an Account of the Origin, Capabilities, and Adaptations of the Numerous Species and Varieties, Native and Foreign, and Especially of the New and Rare Sorts, Suited to Cultivation in the United States. By LUCIUS D. DAVIS. With over 100 illustrations. 8°, $3.50.

"Mr. Davis writes with authority upon his chosen theme. . . . The book is full of information upon the subject of which it treats, and contains many suggestions that may prove helpful."—*N. Y. Times.*

THE WONDERS OF PLANT LIFE

By Mrs. S. B. HERRICK. Fully illustrated. 16°, $1.50.

The only thing aimed at is to give the more important types in a popular way, avoiding technicalities where ordinary language could be substituted, and, where it could not, giving clear explanations of the terms.

"A dainty volume . . . opens up a whole world of fascination . . . full of information."—*Boston Advertiser.*

OUR INSECT FRIENDS AND FOES

How to Collect, Preserve and Study Them. By BELLE S. CRAGIN. With over 250 illustrations. 8°, $1.75.

"Although primarily intended for boys and girls, it can hardly fail to enlist the aid of the older members of the family; and for the amateur collector of all ages who has all the requisite enthusiasm but lacks a practical knowledge of the art of preserving specimens, it should receive a warm welcome."—*N. Y. Commercial Advertiser.*

G. P. PUTNAM'S SONS, 27 & 29 West 23d St., New York

BOOKS FOR THE COUNTRY

AMONG THE MOTHS AND BUTTERFLIES

By JULIA P. BALLARD. Illustrated. 8°, $1.50.

"The book, which is handsomely illustrated, is designed for young readers, relating some of the most curious facts of natural history in a singularly pleasant and instructive manner."—*N. Y. Tribune*

BIRD STUDIES

An account of the Land Birds of Eastern North America. By WILLIAM E. D. SCOTT. With 166 illustrations from original photographs. Quarto, leather back, gilt top, in a box, *net*, $5.00.

"A book of first class importance. . . . Mr. Scott has been a field naturalist for upwards of thirty years, and few persons have a more intimate acquaintance than he with bird life. His work will take high rank for scientific accuracy and we trust it may prove successful."—*London Speaker.*

WILD FLOWERS OF THE NORTHEASTERN STATES

Drawn and carefully described from life, without undue use of scientific nomenclature, by ELLEN MILLER and MARGARET C. WHITING. With 308 illustrations the size of life, and Frontispiece. New edition in smaller form. 8°, *net*, $3.00.

"The authors of this excellent work offer it, not in competition with scientific botanies, but with the hope that by their drawings and descriptions they may make it easy to become acquainted with the wild flowers of the northeastern portion of the United States. Anybody who can read English can use the work and make his identifications, and, in the case of some of the flowers, the drawings alone furnish all that is necessary. . . The descriptions are as good of their kind as the drawings are of theirs."—*N. Y. Times.*

THE SHRUBS OF NORTHEASTERN AMERICA

By CHARLES S. NEWHALL. Fully illustrated. 8°, $1.75.

"This volume is beautifully printed on beautiful paper, and has a list of 116 illustrations calculated to explain the text. It has a mine of precious information, such as is seldom gathered within the covers of such a volume."—*Baltimore Farmer.*

THE VINES OF NORTHEASTERN AMERICA

By CHARLES S. NEWHALL. Fully illustrated. 8°, $1.75.

"The work is that of the true scientist, artistically presented in a popular form to an appreciative class of readers."—*The Churchman.*

THE TREES OF NORTHEASTERN AMERICA

By CHARLES S. NEWHALL. With illustrations made from tracings of the leaves of the various trees. 8°, $1.75.

"We believe this is the most complete and handsome volume of its kind, and on account of its completeness and the readiness with which it imparts information that everybody needs and few possess, it is invaluable."—*Binghamton Republican.*

THE LEAF COLLECTOR'S HANDBOOK AND HERBARIUM

An aid in the preservation and in the classification of specimen leaves of the trees of Northeastern America. By CHARLES S. NEWHALL. Illustrated. 8°, $2.00.

"The idea of the book is so good and so simple as to recommend itself at a glance to everybody who cares to know our trees or to make for any purpose a collection of their leaves."—*N. Y. Critic.*

G. P. PUTNAM'S SONS, 27 & 29 West 23d St., New York

www.ingramcontent.com/pod-product-compliance
Lightning Source LLC
Chambersburg PA
CBHW021958220426
43663CB00007B/869